THE MIRACLE
OF THE CELL

THE MIRACLE
OF THE CELL

MICHAEL DENTON

SEATTLE　　　DISCOVERY INSTITUTE PRESS　　　2020

Description

The Miracle of the Cell provides compelling evidence that long before life emerged on our planet, the design of the carbon-based cell was foreshadowed in the order of nature, in the exquisite fitness of the laws of nature for this foundational unit of all life on Earth. Nowhere is this fitness more apparent than in the properties of the key atomic constituents of the cell. Each of the atoms of life—including carbon, hydrogen, oxygen, and nitrogen, as well as several metal elements—features a suite of unique properties fine-tuned to serve highly specific, indispensable roles in the cell. Moreover, some of these properties are specifically fit for essential roles in the cells of advanced aerobic organisms like ourselves.

Library Cataloging Data

The Miracle of the Cell by Michael Denton

182 pages, 6 x 9 x 0.4 inches & 0.56 lb, 229 x 152 x 10 mm. & 0.25 kg

ISBN-978-1-936599-84-4 (paperback), 978-1-936599-85-1 (Kindle), 978-1-936599-86-8 (EPUB)

Library of Congress Control Number: 2020944212

BISAC: SCI017000 SCIENCE / Life Sciences / Cell Biology

BISAC: SCI007000 SCIENCE / Life Sciences / Biochemistry

BISAC: SCI027000 SCIENCE / Life Sciences / Evolution

BISAC: SCI013000 SCIENCE / Chemistry / General

BISAC: SCI015000 SCIENCE / Cosmology

Publisher Information

Discovery Institute Press, 208 Columbia Street, Seattle, WA 98104

Internet: http://www.discoveryinstitutepress.com/

Published in the United States of America on acid-free paper.

First Edition, September 2020.

CONTENTS

ACKNOWLEDGMENTS

I AM INDEBTED TO SEVERAL KEY TEXTS FROM WHICH MUCH OF THE evidence I cite was taken. These include J. J. R. Fraústo da Silva, *The Biological Chemistry of the Elements* (1991); Wolfgang Kaim, Brigitte Schwederski, and Axel Klein, *Bioinorganic Chemistry*, 2nd ed. (2013); Robert R. Crichton, *Biological Inorganic Chemistry*, 2nd ed. (2012); Rob Phillips et al., *Physical Biology of the Cell*, 2nd ed. (2013); Peter Atkins, *The Periodic Kingdom* (1996); Nick Lane, *The Vital Question* (2015); and Bruce Alberts et al., *Molecular Biology of the Cell*, 4th ed. (2002).

And as always, I am indebted to Lawrence Henderson's great classic *The Fitness of the Environment* (1913).

I also am indebted to Iain Johnston and Tyler Hampton for careful criticisms and readings of earlier drafts of the monograph, for the valuable input of science reviewers later in the process, and for the staff at the Discovery Institute for their considerable editing efforts, especially Jonathan Witt and Rachel Adams.

INTRODUCTION

M Y MAIN AIM IN THE *PRIVILEGED SPECIES* SERIES IS TO PRESENT the evidence that nature is uniquely fit for life as it exists on Earth, not just for the generic carbon-based cell, but also for beings of our biology, and thus to show that the cosmos is not just biocentric but also (no matter how unfashionable it may be in certain quarters) anthropocentric as well.

This particular book is focused on the fitness of nature for the familiar carbon-based cell, which is the basic unit of all life on Earth. In this work I review the properties of many of the atoms of the periodic table, including carbon, hydrogen, oxygen, and nitrogen, as well as phosphorus and several metals, to highlight their unique fitness to serve various biochemical ends in the cell. As the chapters ahead show, the unique fitness of many of the atoms in the first quarter of the periodic table for the design of the cell is stunning. On any consideration, the evidence conveys the irresistible impression that the properties of the atoms have been crafted with incredible precision to perform highly specific functions upon which the life of the cell depends. Together, the collective fitness of the properties of these atoms for the cell make up what I term a "primal blueprint" for the design of the carbon-based cell, a blueprint laid down in the order of things from the beginning of the universe.

Of course, nature's fitness for the cell is not the same as its fitness for human biology. There are many additional elements of fitness in nature which appear to be specifically tailored for beings of our biochemical and physiological design. Some of these have been discussed in *Fire-Maker*, *The Wonder of Water*, and *Children of Light*. But without the astonishing abilities of cells to migrate through the developing embryo following chemical gradients, to change shape, and to morph into various diverse cell types—red blood cells, photoreceptors, epithelial cells, leucocytes, and so forth—no complex multicellular organism would be possible. In effect, the fitness of nature for humans necessitates the prior fitness of nature for the cell. Cells are the crucial stepping stones on the road to humankind.

To keep some focus on our own biology, throughout the text I have highlighted elements of fitness for the cell which are particularly fit for the large complex cells of higher organisms. For the large body cells (more than ten microns across) of mammals, for instance, the diffusion rates of molecules, including oxygen in water, must be very close to what they are. If they were much less, a circulatory system in large multicellular organisms would be impossible, and cells with high metabolic rates like our own would be restricted to tiny bacterial-sized bags of molecules too small to contain an inventory of complex, higher-order molecular systems. Such systems feature microtubules, molecular motors, and other components of the cytoskeleton. These components are vital for abilities that are essential to the embryonic development of advanced multicellular organisms, abilities that include crawling, following chemical gradients, changing shape, and selectively adhering to other cells.

Carl Sagan once remarked, "Extraordinary claims demand extraordinary evidence."[1] The central claim here—that the properties of the atoms reviewed in this book are fine-tuned with stunning precision for the existence of the cell—is indeed extraordinary. But as this book reveals, the evidence for the claim is also extraordinary.

Might there be fine tuning in nature not just for the existence of the cell but also for its origin—for the transition from a lifeless soup of chemicals to a living cell? Or was the first cell engineered by an intelligent agent, as many advocates of intelligent design maintain? I touch on that question in Chapter 8, but in either case, the emergence of primeval life would be *by design*, whether imposed on nature at the origin of the first life, or built into the fabric of nature from the beginning.

A few additional notes on the contents of the book. Chapters 1 and 2 review the historical retreat of old-fashioned vitalism, not only because the retreat is of historical interest but also because it reveals a recurring pattern in the development of biological knowledge, in which a previously inexplicable phenomenon is replaced by the discovery of a special fitness in nature that explains the phenomenon in question without recourse to a vital force. More specifically, this historical lesson may be very relevant to current speculation in the origin-of-life field.

There are some fairly technical sections in this book, particularly in Chapter 6, which describes the fitness of various metal atoms for specific cellular functions, as well as in Chapters 1 and 2, which describe the nature and biological significance of the strong covalent and weak chemical bonds. But my aim has been to write so that a reader with little chemical or biochemical background will still be able to follow the main thrust of the arguments in these sections.

This book offers the most comprehensive review currently in print of the evidence that the laws of nature are fine-tuned for the cell. The same evidence conveys an irresistible impression of design, as much as in any other area of knowledge.

As with all arguments regarding fitness, the evidence builds cumulatively. Chapter 2 reviews the fitness of the carbon atom; Chapter 3, the fitness of the chemical bonds; and Chapter 4, the fitness of carbon's nonmetal atom partners (hydrogen, oxygen, and nitrogen) and water's hydrophobic force (which forges the cell membrane). Chapter 5 reviews nature's fitness for bioenergetics; Chapter 6, the spectacular fitness of so many of the metal atoms for highly specific biochemical functions; and Chapter 7, the unique fitness of that wonder fluid, water.

I hope that any reader who considers the accumulating evidence will, by the end of Chapter 7, be convinced that nature is indeed fine-tuned for carbon-based life, and that the fine tuning conveys an irresistible impression of design.

Lastly, I hope the reader will view the video showing a white blood cell chasing bacteria across a coverslip cited in Chapter 1. Watching it conveys something of the amazing nature of these extraordinary, tiny entities, the basic units of all life on Earth.

1. The Amazing Cell

To see a world in a grain of sand
And a heaven in a wild flower,
Hold infinity in the palm of your hand,
And eternity in an hour.
—William Blake (1803), *Auguries of Innocence*

Cells are amazing. Even to a non-biologist, they convey the impression of being very special objects with extraordinary capabilities. No one who has observed a leucocyte (a white blood cell) purposefully—one might even say single-mindedly—chasing after a bacterium in a blood smear would disagree. To see this in action, watch the brief online video by David Rogers, "Neutrophil Chasing Bacteria."[1] What one witnesses there seems to transcend all our intuitions: A tiny speck of matter, invisible to the naked eye, so small that one hundred of them could be lined up across the top of a pin, is seemingly endowed with intention and agency. It's like watching a house cat chasing a mouse, or a cheetah chasing a gazelle on the African savanna, or indeed a man chasing down a kudu in the Kalahari.

It does not lessen the amazement to conclude that this ability must arise somehow from the atomic complexity that lies within this wondrous speck of matter. For the complexity in which this behavior is instantiated is also far beyond ordinary experience. A cell consists of trillions of atoms, representing the complexity of a jumbo jet and more, packed into a space less than a millionth of the volume of a typical grain of sand. But unlike any jumbo jet, unlike any nano-tech, or indeed unlike even the most advanced human technology of any kind, this wondrous entity *can replicate itself.* Here is an "infinity machine" with seemingly magical powers.

In terms of compressed complexity, cells are without peer in the material world, actualized or imagined. And there is likely far more complexity still to uncover.[2] Even as recently as 1913, when Lawrence Henderson composed his classic *The Fitness of the Environment,* the cell was a black box, its actual molecular complexity a mysterious unknown. Only as the veil began to lift with the mid-century molecular biological revolution did science begin to glimpse the sophistication of these extraordinary pieces of matter. Subsequently, every decade of research has revealed further depths of complexity. The discovery of ever more intricate structures and systems with each increase in knowledge—including vastly complex DNA topologies and a vast and growing inventory of mini-RNA regulator molecules—tells us there is probably much more to uncover. What we glimpse now may be only a tiny fraction of what remains to be discovered.

As Erica Hayden confessed in the journal *Nature,* "As sequencing and other new technologies spew forth data," the complexity unearthed by cell biology "has seemed to grow by orders of magnitude. Delving into it has been like zooming into a Mandelbrot set... that reveals ever more intricate patterns as one peers closer at its boundary."[3]

There is much more to discover about the cell, but even from our current limited knowledge of its depths it is clear that this tiny unit of compact, adaptive sophistication constitutes something like a *third infinity.* Where the cosmos feels infinitely large and the atomic realm infinitely small, the cell feels *infinitely complex.*

But cells are not just complex beyond any sensible measure and beyond any other conceivable material form. They appear in so many ways supremely fit to fulfill their role as the basic unit of biological life. One element of this fitness is manifest in their incomparable diversity of form. Contrast a neuron with a red blood cell, a skin cell with a liver cell, an amoeboid leucocyte with a muscle cell. Each of these different forms is found in the human body, and many more. Or consider the diversity of ciliate protozoans. From the trumpet-like *Stentor* to the dashing *Paramecium*, the universe of ciliate form is absurdly diverse. Or take the radiolarians (see Figure 1.1). Even within this small related group of organisms, the diversity of cell forms is stunning. And yet every member of this fantastic zoo of radiolarian forms is built on exactly the same canonical design.

The unique fitness of the cell to serve as the fundamental unit of life is also manifest in its amazing abilities and the diversity of functions it performs. Even the tiny *E. coli*, a cylinder-shaped bacterium in the human gut, has spectacular capabilities. Howard Berg has marveled at the versatility and capacities of this minuscule organism, calling its talents "legion." He notes that this tiny organism, less than one-millionth of a meter in diameter and two-millionths of a meter long, so small that "20 would fit end-to-end in a single rod cell of the human retina" is nevertheless "adept at counting molecules of specific sugars, amino acids, or dipeptides; at integration of similar or dissimilar sensory inputs over space and time; at comparing counts taken over the recent and not so recent past; at triggering an all-or-nothing response; at swimming in a viscous medium… even pattern formation."[4]

Cells also move in many diverse ways. *E. coli* travel by the propeller-like action of the bacterial flagellum. Others do so via the beating action

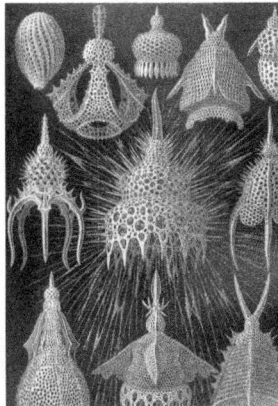

FIGURE 1.1. Radiolarian shells, plate 31 from Ernst Haeckel's *Kunstformen der Natur*, 1904.

of cilia. Some creep and crawl. Some put out pseudopodia and grasp small objects in their immediate vicinity.

Some cells can survive desiccation for hundreds of years. Cells possess internal clocks and can measure the passage of time.[5] They can sense electrical and magnetic fields, and communicate via chemical and electrical signals. Some can encase themselves in armor-like skins. Some may be able to see; one species of ciliate has a lens able to focus an image on another region of the cytoplasm—in effect, an eye. All can replicate themselves with seeming ease, an act far beyond even the most complex human artifact. Some can even reconstruct themselves completely from tiny fractions cut surgically from the cell![6]

These remarkable specks of organized matter have constructed every multicellular organism on Earth, including the human body, itself a vast collective of as many as 100 million million cells. Cells compose the human brain, making a million connections a minute for nine months during gestation. Cells build blue whales, butterflies, birds, and the giant sequoias of Yosemite. Cells constituted the dinosaurs and all past life ever born on Earth. And through the activities of some of the simplest of their kind, cells gradually terraformed the planet over the past 3,000 million years, generating oxygen via photosynthesis and releasing its energizing powers for all the higher life forms. They are the universal constructor set of life on Earth. In short, they can do almost anything, adopt almost any shape, and obey any order. They appear, in every sense, perfectly adapted to their assigned task of creating a biosphere replete with multicellular organisms like ourselves.

When we observe the goings-on of protozoans in a drop of pond water or the antics of an amoeboid leucocyte in the human blood stream chasing a bacterium, it is hard to resist the feeling that these microscopic life forms are sentient, autonomous beings. This was the case when we had relatively primitive microscopic technology more than one hundred years ago,[7] and it is all the more so today.

It is not just their hunting strategies (seen in the video of the leucocyte chasing its prey) that resemble the behaviors of higher organisms.

Another striking example is the courtship rituals of ciliates, rituals that include pre-conjugal mating dances, reciprocal learning, repeated touching of prospective mates, and even deceit and cheating when communicating reproductive fitness to potential mates.[8] One of the founders of behaviorism, Herbert Spencer Jennings, strongly suspected that protozoa were sentient. As he confessed, "If Amoeba were a large animal, so as to come within the everyday experience of human beings, its behavior would at once call forth the attribution to it of states of pleasure and pain, of hunger, desire, and the like, on precisely the same basis as we attribute these things to the dog."[9]

Jennings's thoughts were recently echoed by biologist Brian Ford: "The microscopic world of the single, living cell mirrors our own in so many ways: cells are essentially autonomous, sentient and ingenious. In the lives of single cells we can perceive the roots of our own intelligence."[10]

And as Ford continues, "We regard amoebas as simple and crude. Yet many types of amoeba construct glassy shells by picking up sand grains from the mud in which they live. The typical *Difflugia* shell, for example, is shaped like a vase, and has a remarkable symmetry... We just don't know how this single-celled organism builds its shell."[11]

Even if cells are not sentient beings, their accomplishments, their complexity, their diversity of structure and function, remain to astound us. The unique powers of cells—what Jacques Monod called their "demonic catalytic powers"[12]—and their extraordinary fitness to play their unique role as the building blocks of all life on Earth are a wonder apparent to anyone who gives them even a cursory consideration.

And as we shall see in the chapters ahead, an even greater wonder is the stunning prior fitness in nature that enables the material actualization of the canonical carbon-based cell. This prior fitness is, as we shall see in the chapters ahead, manifest in the unique utility of the properties of a significant number of the atoms in the first half of the periodic table to serve highly specific ends essential for the assembly of the core macromolecular constituents and the physiological functioning of the cell. I call this the unique fitness paradigm.

And as we will also see in later chapters, this prior fitness is manifest also in the extraordinary utility of water to serve as the matrix of the cell, and by chemical processes in the dark vastness of interstellar space that result in the a-biotic synthesis of many of the molecular monomers used by the first cells to build their macromolecular constituents. In other words, the "demonic" fitness of the cell depends on a deeper fitness prefigured into the very fabric of reality. This deeper fitness is inscribed in the laws of nature from the beginning of time, a fitness that reveals the cosmos to be, as Henderson proclaimed, a profoundly biocentric whole.[13]

2. The Chosen Atom

In bodies organised, which true
To type perpetuate themselves.
But through what powers can
Life do this?
Natural or Supernatural?
By properties inherent in
The molecules of which it's made?...
Or by extraneous, imposed power
A *Deus ex Machina* force,
Outside the laws of Nature...
This is the problem: this indeed.
　　　　　　　—Arthur E. Needham, *The Uniqueness*
　　　　　　　　　　　of Biological Materials[1]

IN THE EARLY MODERN PERIOD RIGHT UP TO THE FIRST DECADES OF the nineteenth century, many biologists were vitalists, believing that the unique behavior, characteristics, and abilities of living things that were not shared by non-living things—including sentience, agency, and the capacity for self-replication—were the result of a nonmaterial, indwelling vital spirit. In the seventeenth century Christina, Queen of Sweden, upon hearing René Descartes insist that organisms are analogous to machines, is said to have retorted by saying of a mechanical clock, "See to it that it produces offspring."[2] Christina's challenge has yet to be met. Despite extraordinary advances in nano-technology and supramolecular chemistry, no one has assembled a material entity that can mimic the cell's ability to self-replicate.

In contemplating the astonishing complexity and varied abilities of cells, it is hard to resist the vitalists' inference that cells are endowed with something beyond the ordinary properties of matter. Especially upon observing activities such as the mating behavior of ciliate protozoa, which mimic in intriguing ways the behavior of many birds and mammals, or watching a white blood cell chasing down bacteria in a blood smear, it's hard not to imagine some agency or soul bestowing on them their unique abilities. Such abilities are without peer in any known entity in the inanimate realm, including the domain of our own mechanical creations.

But despite its appeal, the inference that there is something at work in living things beyond the laws at play in the inanimate world has a poor track record. The history of biology testifies that from the dawn of organic chemistry in the early nineteenth century to the discovery of the double helix and the molecular biological revolution in the mid-twentieth century, each major advance in knowledge has led to a retreat of vitalist notions. Each new discovery revealed not a vitalist agency gifting cells with this or that unique behavior or characteristic, but rather some extraordinary prior fitness in the properties of matter. In particular, the properties of many of the atoms of the first half of the periodic table are remarkably fit for the assembly of the cell's chemical constituents and for its physiological functioning.

Vitalism Retreats

As MENTIONED above, in the early nineteenth century, many chemists believed that the unique chemical characteristics of organic compounds derived from some special vital force or agency in the organism. As Lawrence Henderson commented a century later in *The Fitness of the Environment*, "Many organic substances had been separated from the organism, purified and subjected to the usual experiments of the laboratory… But, as Berzelius [one of the leading chemists of that era] believed, a special vital force had presided over their formation and this, therefore, he supposed to be impossible under any other circumstances."[3]

Some chemists at the time imagined the vital agency in very literal terms, like a tiny homunculus in the cell, endowed with the unique ability to assemble atoms into the various complex organic compounds derived from living systems.[4] William Prout, for example, a leading physician and chemist in the early nineteenth century and author of the Eighth Bridgewater Treatise (*Chemistry, Meteorology, and the Function of Digestion*), wrote, "The organic agent... having an apparatus of extreme minuteness, is enabled to operate on each individual molecule separately; and thus, according to the object designed, to exclude some molecules, and to bring others into contact."[5]

Although such views seem archaic today, early nineteenth century belief in this form of vitalism is understandable. No one had synthesized an organic compound in the lab. Compared with inorganic compounds, organic compounds were especially fragile and unstable, decomposing rapidly when removed from the body.[6] Other known differences included their great variety, diversity,[7] and complexity.[8] Isaac Asimov summarized some key differences between the inorganic and organic domains thus:

> Organic materials are much more fragile and easily damaged than inorganic materials. Water (which is inorganic) can be boiled and the resulting steam heated to a thousand degrees without damage. When the steam is cooled down, water is formed again. If olive oil (which is organic) is heated, it will smoke and burn. After that, it will no longer be olive oil.
>
> You can heat salt (which is inorganic) till it melts and becomes red-hot. Cool it again and it is still salt. If sugar (which is organic) is heated, it will give off vapors, then char and turn black. Cooling will not restore its original nature... Organic substances can be treated with heat or by other methods and converted into inorganic substances. There seemed no way [to the chemists of the early nineteenth century], however, of starting with an inorganic substance and converting to an organic substance.[9]

The early nineteenth-century chemists did, therefore, have some justification for believing in a unique biological force. The evidence at the time was consistent with the possibility of some mysterious entity in the cell that assembled atoms into organic compounds. Vitalists such as Prout[10] accepted that atoms combined together in essentially the same way, according to the same rules, in organic substances as in inorganic. But they believed only living systems could assemble them into organic molecules and actualize the remarkable properties of the substances of the organic domain.

But a major plank in the case for vitalism collapsed in 1828 when, in one of the great breakthroughs of nineteenth-century science, a young German chemist, Friedrich Wöhler, synthesized in his laboratory the compound urea, the major constituent of mammalian urine. It was the first time a chemist was able to synthesize a chemical constituent of a living organism from simple, inorganic compounds. Its synthesis required no vital force and its atomic constituents were combined together in exactly the same way as they would be in an ordinary inorganic compound.

Wöhler obtained urea ($CO(NH_2)_2$) by treating inorganic silver cyanate ($AgOCN$) with another inorganic compound, ammonium chloride (NH_4Cl). He wrote triumphantly to his mentor Berzelius, a leading vitalist, "I must tell you that I can make urea without the use of kidneys of any animal, be it man or dog. Ammonium cyanate is urea."[11]

As Frances Preston Venable sums up:

It was Wöhler's brilliant synthesis of urea which finally broke down this barrier, proving the forerunner of many syntheses, and inciting numbers of workers to labor in this lucrative field. It is true that the synthesis had not been made directly out of the elements; but still it was out of substances then regarded as inorganic that he had prepared one of the most interesting and best known of animal products. Of course the dying away of the old belief was slow, but Wöhler's discovery is commonly pointed to as marking the beginning of organic chemistry as a science.[12]

In 1845, Hermann Kolbe put another nail in the coffin by synthesizing the organic substance acetic acid from carbon disulfide in the lab.[13] After Kolbe's synthesis, the dam broke, and scientists synthesized more and more organic compounds in the lab. It became clear that at least the basic compounds of living things could be made without some vital agency in the cell.

As knowledge of the chemicals of life grew throughout the nineteenth century, it became ever clearer that not only were the chemical constituents of living things perfectly natural compounds, but also that the carbon atom in conjunction with hydrogen, oxygen, and nitrogen (which make up the bulk of organic substances) possessed a special chemical fitness for the assembly of a vast inventory of complex and diverse organic compounds (acids, sugars, ethers, esters, alcohols) necessary to build complex biochemical systems. As Henderson commented, "The compounds of organic chemistry gradually came to be recognized as different from inorganic substances only in the special characteristics of the elements carbon, hydrogen, and oxygen when in chemical union with one another, just as the compounds of any other elements have their own specific characteristics."[14]

Today, a century after Henderson, still no other chemistry is known that can provide such a cornucopia of chemical compounds from which to choose a set of building blocks for a living system and supply all its necessary metabolites. And curiously, the vitalist notion that there was some fundamental difference between the chemicals of life and those of the inanimate realm was retained, even emphasized. But the difference was no longer attributed to a vital supernatural artificer in the cell. It was instead attributed to the unique emergent and natural chemical and physical properties of the carbon atom in combination with hydrogen (H), oxygen (O), and nitrogen (N). Rather than a wonder-working agency in the cell endowing life's compounds with their unique characteristics, the real wonder-worker was now seen as the unique chemical fitness of certain atoms in the periodic table for life. Agency had been replaced by chemical *fitness*. A wonder external to nature had been replaced by a wonder immanent in the properties of matter—vitalism by fitness, immediate design by ultimate design, immediate agency by ultimate agency.

It is important to note, however, that although the classic form of vitalism, which postulated a vital agency in the cell to account for life's chemistry, was abandoned in the nineteenth century as the wonder of carbon chemistry became increasingly apparent, there are still today biological phenomena which are beyond any explanation in terms of the current laws of physics and chemistry. An obvious example is the realm of sentience, mind, and consciousness.[15] And whether or not there are laws of nature which apply uniquely to the organic realm remains an open question that we can only hope will be answered by future advances in science.

But whatever future advances of science might reveal regarding the various causal factors at work in living systems, the core claim defended in the chapters ahead is based on an assessment of the scientifically established properties of matter and laws of nature accepted by all biologists today. That there is a profound prior fitness in nature which enabled the actualization of the canonical cell as it exists on Earth is a claim independent of whatever might have been the direct causal factors responsible for the assembly of the first living cell on Earth or the exact physical and chemical steps by which the miracle was actualized. Whether these factors were Darwinian, Lamarckian, vitalistic, or some other is a debate of great interest but largely peripheral to the focus of this book.

The Infinite Inventory

THE DEVELOPMENT of organic chemistry is one of the great episodes in the history of science, and was described by Henderson as "one of the greatest achievements of the nineteenth century."[16] Others concur. Jan Mulder entitled a paper reviewing its development as "Looking Back in Wonder."[17] Many other authors, including Asimov[18] and Alfred Russel Wallace, co-founder with Charles Darwin of the theory of evolution by natural selection, have waxed lyrical about the wondrous universe of carbon chemistry.

By the beginning of the twentieth century, more than 100,000 organic compounds had been documented.[19] And all the basic compounds of living organisms—the twenty common amino acids used in proteins and the four nucleotides used in DNA, as well as many of the sugars and fats and fatty acids found in living organisms—had been synthesized in the lab.

Of all the elements, carbon stands alone in its ability to form a vast array of complex organic compounds with diverse chemical and physical properties. Indeed, the number of known carbon compounds is currently estimated to be close to ten million, greater than the total of all other non-carbon compounds combined and much larger than Henderson's estimate from a century ago.

And aside from molecules that include carbon, there are many molecules that contain only carbon. Carbon makes up substances as diverse as coal, diamond (the hardest mineral known), and graphite (one of the softest), as well as complex structures such as fullerenes and nano-tubes. In recent decades chemists announced the discovery of another carbon compound, graphene, which consists of a flat monolayer of carbon atoms packed tightly into a two-dimensional honeycomb arrangement. Its most remarkable characteristic is its strength: it is one hundred times stronger than an equivalent monolayer of steel. Graphene conducts electricity as well as copper does, and conducts heat better than can any known material.

However, the diversity of chemical forms that can be assembled using carbon alone pales against the fantastic diversity of compounds that can be assembled when carbon combines with other atoms.

Carbon and hydrogen combinations form the universe of hydrocarbons. Some hydrocarbons are long, chain-like molecules, such as pentane and butene. Others contain cyclic or ring-like formations, such as benzene. And it is not just the number of chemical structures that dazzles, but also the variety and diversity of properties. Plastic milk jugs, DVD discs, oils, petroleum, kerosene, and naphthalene (moth balls) are all combinations of carbon and hydrogen atoms.

Combining carbon with both hydrogen and oxygen opens another universe of compounds, including alcohols such as ethanol and propa-

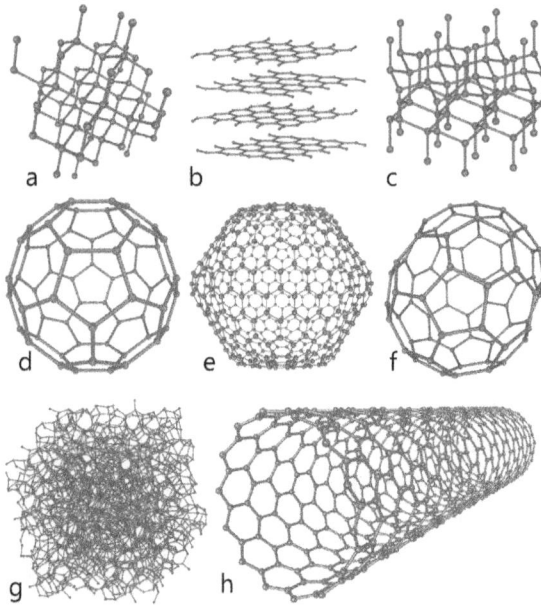

Figure 2.1. A few structures composed of the carbon atom: a) diamond; b) graphite; c) lonsdaleite; d–f) fullerenes; g) amorphous carbon; h) carbon nanotube.

nol, aldehydes, ketones, and the carboxylic acids. This combination also creates the vast variety of fatty acids, composed of a long hydrocarbon chain that is attached to a carboxylic acid group at one end. Carbon, hydrogen, and oxygen are also responsible for the sugars, including glucose and fructose. Beyond that, this triad creates cellulose (the hard substance of wood), beeswax, vinegar, and formic acid. All of these belong to this group of carbon compounds.

Throwing nitrogen into the mix leads to a further multiplicity of compounds, including the building blocks of proteins: amino acids. It also creates a set of cyclic compounds known as the nitrogenous bases, some of which are important building blocks of DNA. This combination is found in items as diverse as dyes, antibiotics, explosives, caffeine, and urine.

By the late nineteenth century and the beginning of the twentieth, when the full wonder and uniqueness of carbon's lavish inventory had been established, Alfred Russel Wallace described the sheer number and diversity of the denizens of this unique domain in both of his works on natural theology, *Man's Place in the Universe* and *The World of Life: A Manifestation of Creative Power, Directive Mind and Ultimate Purpose*. In the first of these, he wrote:

> The chemical compounds of carbon are far more numerous than those of all the other chemical elements combined... And the marvel is still further increased when we consider that the innumerable diverse substances produced by plants and animals are all formed out of the same three or four elements. Such are the endless variety of organic acids, from prussic acid to those of the various fruits; the many kinds of sugars, gums, and starches; the number of different kinds of oil, wax, etc.; the variety of essential oils which are mostly forms of turpentines, with such substances as camphor, resins, caoutchouc [natural rubber], and gutta-percha; and the extensive series of vegetable alkaloids, such as nicotine from tobacco, morphine from opium, strychnine, curarine, and other poisons; quinine, belladonna, and similar medicinal alkaloids... all alike consisting solely of the four common elements from which almost our whole organism is built up. If this were not indisputably proved, it would scarcely be credited.[20]

The total number and diversity of possible chemical structures that may be constructed out of carbon, oxygen, hydrogen, and nitrogen staggers the imagination. Together, these elements form what is in effect a universal chemical constructor kit ideally suited for the construction of the myriads of chemical compounds the cell employs. The need for such a vast inventory of organic compounds is indicated to a degree by published metabolic pathway charts. The charts show the maze of chemical pathways and the huge number of different compounds which undergo chemical transformations in the course of metabolism in a typical cell.

King Carbon

GIVEN THE unparalleled fecundity of the carbon universe, it is somewhat curious that in many ways carbon is not a very prepossessing substance. Think of a lump of coal or a piece of graphite or a heap of soot. Diamond has allure, but on the whole, in most of its elemental forms, carbon does not impress. Its relative un-reactivity[21] adds to the impression that it is mundane compared with other more spectacular and reactive atoms like sodium or oxygen. As chemist Peter Atkins comments, carbon seems in terms of reactivity a "particularly mediocre" atom and "easygoing in the liaisons it forms."[22]

But carbon does have chemical fecundity, which elevates it into a category all its own, creating the vast array of chemical combinations described above. For this reason, Atkins termed carbon "the King of the Periodic Kingdom."[23] Certainly without the vast inventory of complex molecules of utterly diverse chemical properties gifted to us via the unique properties of this king of atoms, there would be no organic plenitude to satisfy the complex metabolic needs of the cell. In all probability, there would be no chemical life in the universe. Atkins goes so far as to say that the "property we term 'life' stems almost in its entirety" from the region of the kingdom containing carbon.[24]

Carbon-Carbon Bonds: Of all the atoms of the periodic kingdom, including carbon's three associates which make up the substance of organic compounds—hydrogen (H), oxygen (O), and nitrogen (N), only carbon can bond firmly with itself to form chains of atoms (i.e., C-C-C-C) of almost unlimited length. In this ability, carbon is unique. No other atom in ambient conditions, not oxygen, nitrogen, hydrogen, or silicon, possesses this ability to anything like the same degree as carbon.

More than anything, it is the stability of carbon-carbon bonding that enables organic compounds to grow to almost unlimited size and complexity. In the case of organic molecules containing carbon, as Asimov put it, "Carbon atoms can join one another to form long chains or numerous rings and then join with other kinds of atoms as well. Very large molecules may be formed in this way without becoming too rickety to exist. It is not at all unusual for an organic molecule to contain a million atoms."[25] Large molecules even remotely as complex as proteins or other macromolecules are simply unknown outside the domain of organic chemistry. Many authors have stressed this. As Primo Levi puts it, "Carbon, in fact, is a singular element: it is the only element that can bind itself in long stable chains without a great expense of energy, and for life on earth (the only one we know so far) precisely long chains are required. Therefore carbon is the key element of living substance."[26]

Tetravalency: Carbon possesses another element of fitness: it is *tetravalent*, meaning it can form four electron-sharing chemical bonds with other atoms including itself. By way of comparison, nitrogen can only form three such bonds with other atoms, oxygen two bonds, and hydrogen one. As Arthur Needham comments, "The four bonds of each carbon atom are directed towards the corners of an imaginary tetrahedron, with the carbon atom as its centre, so that by bonding with other carbon atoms an indefinitely extended three-dimensional fabric is possible, similar to that of water... The basis of bioplasm is essentially a fabric of this kind."[27]

Its tetravalency further contributes to carbon's unique fitness for the formation of the vast inventory of organic molecules.

Multiple bonds: A third element of fitness which contributes to the limitless fecundity of the organic universe is carbon's ability to form multiple, stable bonds with itself and other atoms. This is a consequence of carbon's relatively small atomic radius, which means the bond distances are short and thus relatively strong.[28] The other small, nonmetal atoms in period two, including two of carbon's partners, oxygen and nitrogen, also share this capacity.

Carbon can form single, double, and triple bonds with other atoms, including itself. Nitrogen can form single, double, and triple bonds, and oxygen can form single and double bonds. The nonmetals right below them in the periodic table (silicon, phosphorus, and sulfur) form such bonds less readily because their larger atomic radii render multiple bonds less stable.[29]

The Right Strength: The strength of the bonds that link the carbon atoms with other carbon atoms and with the other atoms of organic chemistry—mainly hydrogen, oxygen, and nitrogen—has to be commensurate with their chemical manipulation by the molecular machinery in the cell; otherwise the above ensemble of fitness would be to no avail. Fortunately, the strength of these chemical bonds, and their energy levels—which play a major role in determining the strength of the bonds—are just what they need to be.

To understand why the energy levels of the chemical bonds in organic compounds are indeed in the right range for biochemical manipulations, consider briefly how the cell's molecular machines carry out chemical reactions. Basically, a combination of two factors are involved. One involves using the energy of molecular collisions to weaken chemical bonds, and the other, specific conformational movements in an enzyme molecule, which strain a particular bond in a particular substrate molecule, lowering the energy level of the bond. In chemical jargon, it decreases the activation barrier, making the bond weaker and easier to break.[30]

The need for the decrease in the activation barrier is real, because at ambient temperatures the energy imparted by molecular collisions is insufficient to overcome the energy barriers of most organic bonds.[31] (And this is why the organic compounds which make up the substances of the body remain chemically stable for relatively long periods of time). By reducing the activation barriers, many more molecular collisions have sufficient energy to break the bonds. Although protein conformational changes involve energy levels significantly less than those of a covalent bond,[32] about one tenth, fortunately they are still sufficient in the ambient temperature range to significantly strain particular bonds to lower their activation energy to levels that can be broken by less energetic but more frequent molecular collisions.

If organic bonds were substantially stronger in the ambient temperature range, say as strong as in many inorganic compounds which may be two to three times as strong[33] (and which can only be broken by heating to very high temperatures), protein movements could not significantly weaken particular bonds, i.e., decrease the activation barrier for particular reactions. Consequently, the sorts of controlled chemical reactions carried out in living cells would be greatly constrained. Moreover, not only would proteins be unable to exert sufficient conformational strain to significantly weaken particular bonds, but molecular collisions in the ambient temperature range would only very rarely impart sufficient energy to overcome energy barriers and cause bonds to break. On the other hand, if organic bonds were substantially weaker in the ambient temperature range, disruption via molecular collisions would dominate and no controlled chemistry would be possible.

It turns out that the actual strength of organic bonds, as with so many other examples of the fitness of nature for life, is situated in a Goldilocks zone, neither too strong nor too weak, but just right. If the bonds were stronger or weaker by a single order of magnitude, the controlled chemistry of the cell would very likely be impossible. And it is surely an arresting fact, testimony to the prior fitness of nature for carbon-based life, that this Goldilocks zone represents an inconceivably tiny band in the vast spread of energy levels in the cosmos. (Gravity, for example, is at least 10^{36} times weaker than the strong nuclear force.[34])

In short, biochemistry is only possible because carbon compounds in the ambient temperature range are, as described by Needham, uniquely "metastable."[35] As he points out, while carbon compounds are relatively stable in this temperature range and can persist without undergoing chemical change for long periods in the cell, they are "notable for lability as much as for their stability. Few remain unchanged when heated above 300°C, and most are gaseous at that temperature, if not already decomposed. As in so many other respects, carbon seems to have the best of both worlds, in fact, combining stability with lability, momentum with inertia."[36] As Henderson put it in *Fitness*:

> Not less valuable for the organism than the multiplicity of organic substances, and the diversity of their properties, are the great variety of chemical changes which they can undergo, and that characteristic instability which renders such great complexity of chemical behavior easily attainable. In short, organic substances are uniquely fitted not only to provide complexity of structure to the organism, but also, through their instability and manifold transformations to endow it with diverse chemical activities, with complexity of physiological function.[37]

Similar Affinity: Many authors also have stressed another characteristic of carbon bonds: the energy levels of carbon bonds do not differ much from one partner element to the next. As N. V. Sidgwick notes in his classic *The Chemical Elements and Their Compounds*, "The affinity of carbon for the most diverse elements, and especially for itself, for hydrogen, nitrogen, oxygen, and the halogens, does not differ very greatly: so that even the most diverse derivatives need not vary much in energy content, that is, thermodynamic stability."[38] Robert E. D. Clark expanded on the same point, writing that carbon "is a friend of all. Its bond energies with hydrogen, chlorine, nitrogen, oxygen or even another carbon differ little. No other atom is like it."[39] Kevin W. Plaxco and Michael Gross concur, and elaborate in their well-known *Astrobiology*:

> Carbon presents a fairly level playing field in which nature can shuffle around carbon-carbon, carbon-nitrogen, and carbon-oxygen single and double bonds without paying too great a cost to convert any one of these into another... Given all this, it's no wonder that on the order of ten million unique carbon compounds have been described by chemists, which is as many as all of the described non-carbon-containing compounds put together.[40]

The Right Temperature Range: The necessity for metastable organic compounds to enable the controlled chemistry in the cell has a further intriguing consequence; it constrains the life-friendly temperature range to an extraordinarily narrow band within the immensity of the total range of temperatures in the cosmos.

The upper temperature limit for life is not much above 100°C.[41] This is because of the characteristic instability of most organics as temperatures rise beyond that point. Stanley Miller and Leslie E. Orgel noted this in their book *The Origins of Life on the Earth*.[42] The key amino acid alanine, for example, has a half-life of 20 billion years at 0°C, three billion years at 25°C, but only ten years at 150°C, a decrease of more than a billion-fold. And alanine is not exceptional.[43] As I pointed out in *Nature's Destiny*, "Many vitamins, including vitamin C, folic acid, and some of the other B vitamins—B1 and B6, for example—are rapidly broken down above 100°C."[44] One report in the journal *Nature*[45] showed that the half-life of many of the key organic compounds used by living things—including the amino acids used in proteins, the bases used in DNA, and the adenosine triphosphate (ATP) used for energy metabolism in cells—decompose at rates too fast to measure, or have half-lives on the order of minutes or seconds, at 250°C.[46]

The lower level for controlled biochemistry has not been ascertained. However, it is known that some organisms can function at temperatures as low as -20°C, below which cell vitrification causes metabolism to cease.[47] Whether life could exist at even lower temperatures is not known because no detailed studies have been carried out on the biochemistry of cells at temperatures below -20°C in fluids that are liquid at very low subzero temperatures. But being generous and allowing for a "slow" biochemistry at -50°C (in some fluid other than water) and an upper limit of, say, 130°C, the temperature range fit for biochemistry would still occupy only an infinitesimal fraction of the vast range of temperatures in the universe.[48]

That by itself is striking. But there's something else. This temperature range just so happens to be almost the same as the temperature range in which water is a liquid in ambient conditions on Earth,[49] surely one of the most extraordinary and consequential bio-friendly coincidences in nature. For if these two independent ranges didn't happen to

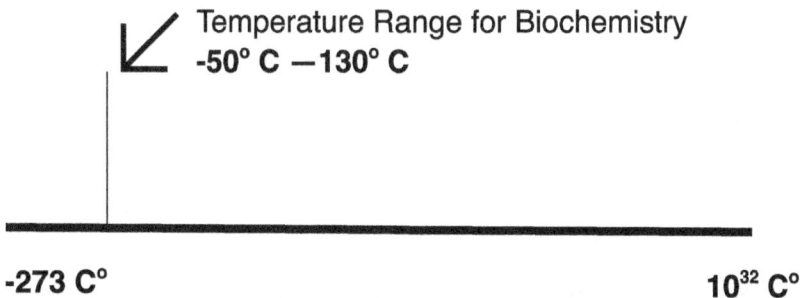

Temperature Range for Biochemistry
-50° C — 130° C

-273 C° 10^{32} C°

Figure 2.2. Cosmic temperatures, from the temperature of absolute zero to the temperature of the Big Bang.

overlap, there would be, in all probability, no carbon-based life on Earth or indeed anywhere in the universe.

Multiple Fitness

IN SUM, carbon is fit in many different ways for the assembly of the complex molecules of life:

1. It forms stable bonds with itself.
2. It forms up to four bonds, being tetravalent.
3. It forms multiple bonds with itself and other atoms.
4. The energy levels of carbon bonds are just right for biochemical manipulation in the ambient temperature range—not too strong and not too weak—described as being "metastable."
5. The energy levels of the covalent bonds that carbon forms with its other nonmetal partners in organic compounds are similar.
6. The metastability of carbon compounds is in the same temperature range that water is a liquid.

And there is yet another element of fitness which will be discussed in the next chapter: the unique directional nature of the bonds carbon forms with other atoms in organic compounds, which, we will see, plays a vital role in the assembly of complex macromolecules of defined 3-D shapes.

That the carbon atom is uniquely fit for the chemistry of life is not the view of an esoteric minority of researchers or of any special pleading. The peerless fitness of the carbon atom to build a universe of diverse chemicals and fantastically complex macromolecules like proteins and DNA has been recognized by the majority of authors and researchers cognizant of the facts. This has been the case for more than a century.[50]

By the end of the nineteenth century, all attempts to account for the very remarkable chemical properties of the organic realm in terms of a mysterious "vital force" had been abandoned. Even advocates of design such as Alfred Russel Wallace saw chemical design as inherent in nature and sufficient without augmentation by some mystical vitalistic agency.

While there may be other forms of chemical life, perhaps based on boron or silicon (for which there is at present no empirical evidence), what seems not in doubt is the supreme fitness of carbon for any chemical life form analogous to that on Earth. As Gross and Plaxco confess in *Astrobiology*, "In the end there may very well be only a single element—carbon... the basis of all life on Earth—that is able to support the complex chemistry presumably required to create a self-replicating chemical system."[51]

Finally, what is particularly striking about the properties of the carbon atom is that they appear to be fine-tuned in several different but complementary ways to generate the plenitude of compounds uniquely useful to life. Such a suite of properties, all seemingly arranged to generate a vast inventory of molecules ideal for the biochemistry of living cells, conveys a powerful impression of contrivance. More than a century ago Wallace expressed, in his *World of Life*:

> We see, therefore, that carbon is perhaps the most unique, in its physical and chemical properties, of the whole series of the elements, and so far as the evidence points, it seems to exist for the one purpose of rendering the development of organized life a possibility. It further appears that its unique chemical properties, in combination with those of the other elements which constitute protoplasm, have enabled the various forms of life to produce that almost infinite variety of substances adapted for man's use and enjoyment, and especially to serve the purposes of his ever-advancing research into the secrets of the universe.[52]

It turns out that Wallace, peering into the biochemical basis of life, saw what has now become even clearer. Many have believed (and many still do believe) that Darwin drove teleology out of biology forever. But more than a century and half of scientific research since Darwin has shown that the fitness of nature for life on Earth, exemplified so wonderfully in the chosen atom, points irresistibly to purpose and design.

3. The Double Helix

Twenty angstrom units in diameter, seventy-nine billionths of an inch. Two chains twining coaxially, clockwise, one up the other down, a complete turn of the screw in 34 angstroms. The bases flat in their pairs in the middle, 3.4 angstroms and a tenth of a revolution separating a pair from the one above or below. The chains held by the pairing closer to each other around the circumference one way than the other, by an eighth of a turn, one groove up the outside narrow, and other wide. A melody for the eye of the intellect, with not a note wasted.
—Horace Judson, *The Eighth Day of Creation*[1]

Carbon's unique chemistry, manifest in the sheer number and diversity of its compounds, is ideally fit to provide the cell with a fantastic inventory of small molecular building blocks—sugars, amino acids, nucleotide bases, fats, steroids, and so forth. Yet this abundance cannot by itself explain all of the seemingly miraculous abilities of cells, such as how enzymatic catalysis is carried out or how cells transmit genetic information to their two daughter cells in cell division. These phenomena were still complete mysteries in the early twentieth century. And for many researchers at the time, it seemed these abilities might not be explicable in terms of the laws of physics and chemistry that applied in the inorganic world or the laboratory. Consequently, vitalist notions still found some support among many biologists.

James Watson commented in *Molecular Biology of the Gene*:

Through the first quarter of this century, a strong feeling existed in many biological and chemical laboratories that some vital force outside the laws of chemistry differentiated between the animate and the inanimate. Part of the reason for the persistence of this "vitalism" was that the success of the biologically orientated chemists (now usually called biochemists) was limited. Although the techniques of the organic chemists were sufficient to work out the structures of relatively small molecules like glucose... there was increasing awareness that many of the most important molecules in the cell were very large—the so-called macromolecules—too large to be pursued by even the best of organic chemists.[2]

And, as Watson further noted, even the demonstration by James B. Sumner in 1926 that the enzyme urease was a protein and could be crystallized in the lab "did not dispel the general aura of mystery about proteins."[3] These were still undecipherable by the techniques available at the time, so "it was still possible, as late as 1940, for some chemists to believe that these molecules would eventually be shown to have features unique to living systems."[4] The elusive genetic material was widely considered to be made up of proteins (not DNA), which were known to be associated with chromosomes in the cell nucleus. DNA molecules were also known to be a constituent of the chromosomes, but "these were thought to be relatively small and incapable of carrying sufficient genetic information."[5] And as Watson concludes, "The feeling was often expressed that something unique about the three-dimensional organization of the cell gave it its living feature... that some new natural laws, as important as the cell theory or the theory of evolution, would have to be discovered before the essence of life could be understood."[6]

Max Delbrück, one of the founding fathers of molecular biology and leader of the celebrated Phage Group[7] (whose work led to the discovery that bacteriophage DNA carries genetic information and specifies the assembly of phage capsids in the bacterial cell), commented to Horace Judson during a conversation that Judson reported in his classic *The Eighth Day of Creation*:

> Both proteins and nucleic acids were hopelessly inadequately characterized, in those days. Proteins were characterized a little more, because we knew that there were twenty essential amino acids—or twenty-*odd*

amino acids: one didn't, after all, know really how many. But whether proteins were built in a regular repetitive way or in a very specific way was still very unknown in the forties. And similarly, DNA—one knew it was a fibrous molecule, one didn't really know how the nucleotides went in together, at this time, or even whether it had branches in it or not.[8]

Ideas on the three-dimensional structure of proteins were no less ill defined. Scientists agreed that proteins were big molecules, or macromolecules, probably composed of linear chains of amino acids, but didn't know how their primary sequences were determined, how they were synthesized, and how they arrived at their native three-dimensional conformations.[9] So little was known about proteins or DNA in the 1930s and '40s that, as mentioned above, most biochemists thought that proteins, rather than nucleic acids, played the primary role in heredity and made up the material structure of the gene.[10]

In short, there was every reason, as in the early nineteenth century, for vitalistic speculation. As Delbrück pointed out in the 1930s, "Genes at that time were algebraic units of the combinatorial science of genetics, and it was anything but clear that these units were molecules analyzable in terms of structural chemistry. They could have turned out to be submicroscopic steady-state systems, or they could have turned out to be something unanalyzable in terms of chemistry."[11]

As Francis Crick recalls about the years leading up to the molecular biological revolution:

> Looking back, what was really striking about that period was not, as we take it now rather obviously, that chemistry and physical chemistry, and the physics associated with those subjects, are essential to understanding biology at the molecular level. Not everybody agreed with that, in particular Max Delbrück, who was a physicist who went into biology.... He was hoping to discover new laws of physics which emerged when he looked at these extremely mysterious biological processes of how you get replication, which in those days seemed utterly baffling.[12]

Erwin Schrödinger also alluded to Delbrück's quasi-vitalist hope that there might be special laws of physics to account for heredity: "From Delbrück's general picture of the hereditary substance it emerges that living matter, while not eluding the 'laws of physics' as established up to date, is likely to involve 'other laws of physics' hitherto unknown, which, however, once they have been revealed, will form just as integral part of this science as the former."[13]

In sum, despite the fitness of the carbon atom for the diverse organics required for the cell, this knowledge left unsettled how higher-order phenomena like enzymatic catalysis and heredity were carried out in the cell. There was a gap, and it was not clear how it might be closed.

After Watson and Crick

BUT DELBRÜCK'S hope that new laws might be discovered was not to be. The mid-twentieth-century molecular biological revolution revealed that no laws unique to biology were necessary to account for enzymatic catalysis or the phenomenon of heredity.

It turned out that what biochemist Jacques Monod referred to as the "demoniacal functions"[14] of enzymes as well as the storage and replication of genetic information could be largely explained by the same laws of physics and chemistry that applied in the inanimate realm; that something as mundane as molecular shape—the exact positioning of atoms in space—could achieve what was previously attributed to something beyond ordinary chemistry.

During the 1940s and 1950s, knowledge of the molecular structure of the core macromolecules of life increased dramatically due to a series of revolutionary advances brought about by the application of powerful new technologies such as X-ray crystallography and electron microscopy. The advances finally clarified the basic structure and function of the key macromolecules in the cell.

And they revealed something quite astonishing: the atomic constituents of the complex macromolecules in the cell could be arranged in unique, highly specific 3-D conformations, and it was this that enabled

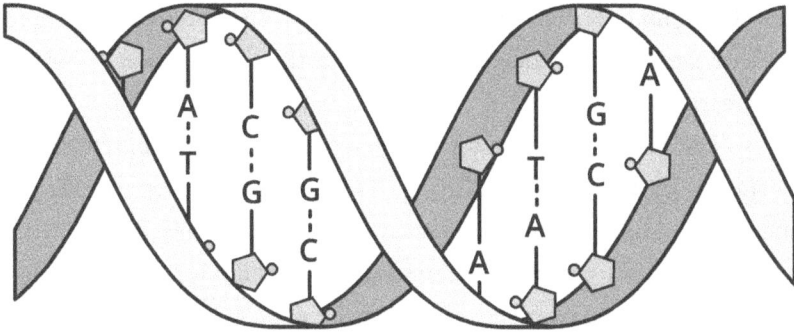

Figure 3.1. DNA's double helix. The twisting ribbons are the phosphate chains. The letters are the pairings of nucleotide bases—adenine (A) and thymine (T), and cytosine (C) and guanine (G). The pentagons attached to the letters are the sugars.

macromolecules such as proteins and DNA to carry out highly specific biochemical, enzymatic, and genetic functions previously attributed to mysterious vital forces.

The protein structures brought to light by X-ray crystallography were the stuff of science fiction. These were objects consisting of up to thousands of atoms arranged into highly specific 3-D arrangements, stranger and more beautiful and more dauntingly complex than any previously conceived aggregate of matter. Max Perutz, quoting J. C. Kendrew's earlier paper, commented on the 3-D conformation of the atoms in myoglobin, the first protein to have its structure determined in atomic detail: "Perhaps the most remarkable features of the molecule are its complexity and its lack of symmetry. The arrangement seems to be almost totally lacking in the kind of regularities which one instinctively anticipates, and it is more complicated than has been predicated by any theory of protein structure."[15]

Rosalind Franklin's X-ray diffraction work carried out at my alma mater, King's College in London, led to the discovery that the atoms in DNA are deployed in the form of a double helix. Although the double helix is a far more regular structure than the bewildering forest of atoms in proteins such as myoglobin or cytochrome, the structure nonetheless revealed again how molecules held in specific 3-D conformations could perform very specific functions, in this case the transmission of genetic information.

In the 1960s, shortly after the glory days of the double helix, my doctoral supervisor at King's College, Henry Arnstein, often referred to the shock he and many other biochemists felt when it became clear that all the atoms in a complex protein could be maintained in precise spatial conformations, and that this specificity of atomic arrangement was key to enabling matter to carry out the seemingly miraculous chemical activities within the cell. A veil had been drawn aside. This was the molecular secret of life.

Judson, in an insightful foreword to the first edition of *The Eighth Day of Creation*, emphasized that a deeper understanding of high specificity was key to ushering in the revolution:

> In the transformation of biology, the great underlying shift of view was the development of the concept of biological specificity. In the mid-thirties, biologists and biochemists certainly spoke of specificity. They had to do so, for many of the phenomena they dealt with—genes (whatever they were in substance), enzymes and antibodies (known to be protein)—were highly specific in action. Yet specificity was really a term almost empty of meaning.... Forty years later, biological specificity is richly stuffed with meaning.[16]

Although the discovery of the structure of DNA was one of the seminal discoveries of twentieth-century science, the double helix is certainly not the secret of life that Crick claimed it to be in 1953. Nor is any other molecule. Indeed, it is possible to conceive of carbon-based life without DNA, one that uses another self-replicating genetic polymer, perhaps closely related to DNA but built from different nucleotides or perhaps a very different type of polymer built from quite different basic units. A group of researchers at the University of Florida and another group in Oxford working in the field of artificial life are now busy redesigning the genetic code[17] and building analogues of DNA.[18] It may be that these alternatives are inferior in one way or another to DNA;

nonetheless, DNA may turn out to be just one of a number of complex polymers capable of performing the genetic function in carbon-based life forms.

What is unarguable, however, is that the functions of all the macro-molecules in current biological systems on Earth depend critically on the ability to deploy multiple atoms (sometimes thousands) in very specific irregular spatial conformations. And one can assume that even artificial life, if we ever invent it—and alien life, if it exists—also will depend on highly specific 3-D molecular conformations of their chemical components.

No chemical life that we can conceive of (and many definitions are given in the literature[19]) would be feasible without complex molecular machines that can carry out defined tasks. And any sort of molecular machines that can carry out specific biochemical functions would necessarily depend on highly precise and stable 3-D arrangements of atoms. For instance, enzymes catalyze life-essential processes by binding to specific substrates, increasing the rate of conversion to an end product by thousands, or even millions, of times per second.[20] No enzyme could manage any such task unless the atoms around the active site were deployed in very exact spatial arrangements to bind the substrate.

Strong Bonds

THE ARRANGEMENT of atoms in complex bio-macromolecules into highly specific 3-D conformations depends on two types of chemical bonds, strong or covalent bonds (discussed in the previous chapter) and another set of quite different bonds, what are called weak bonds.

Covalent bonds form when atoms share electrons to complete their outer electron shells. Familiar compounds in which the atoms are linked by covalent bonds are carbon dioxide (CO_2), water (H_2O), ammonia (NH_3), and methane (CH_4) or marsh gas, shown in Figure 3.2. (For simplicity's sake I'm describing the Lewis Theory for chemical bonds, which is precise enough for our purposes here. However, the most recent scien-

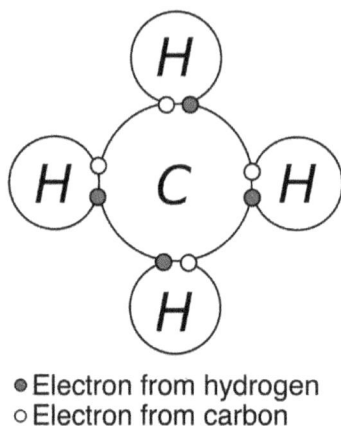

● Electron from hydrogen
○ Electron from carbon

Figure 3.2. Covalent bonds in methane.

tific view, which gives a more detailed explanation of chemical bonds, is the Molecular Orbital Theory.)

In organic compounds, all the bonds between the constituent atoms, such as C-H, C-O, C-N, and N-O bonds, are strong covalent electron-sharing bonds. The crucial feature of covalent bonds is that they are spatially constrained by the existence of other bonds in the molecule. In other words, the bonds are *directional*. As Peter Atkins explains about covalent bonding, "The ability of an atom partially to release electrons to form a covalent bond in one direction will affect its ability to release them in a different direction. As a result, the arrangement of atoms in a molecule has a fixed, characteristic geometry... covalent compounds... are discrete, often small groupings of atoms... with characteristic shapes."[21]

The fact that the bonds in the molecular building blocks of the cell's key macromolecules are directional and spatially constrained is of very great consequence. Why? Because a complex macromolecule in which all the atoms must be deployed in stable, specific spatial arrangements, to serve particular biological functions, cannot be assembled from subunits in which the bonds are not directional and spatially constrained.

Ionic bonds, the other type of strong chemical bond, are also strong, but they are not directional and spatially constrained. Ionic bonds involve the donation or acceptance of electrons between two atoms, leaving one atom (the electron donor) positively charged and the other atom (the electron acceptor) negatively charged. Thus, in the formation of salt (NaCl), the sodium atom donates an electron to a chlorine atom, conferring a positive charge on the sodium atom (Na^+) and a negative charge on the chlorine ($Cl–$). Ionic bonds bind the atoms of most inorganic compounds in the mineral world.[22] But although ionic bonds are strong—generally more than twice the energy of the covalent bonds that carbon makes with hydrogen, oxygen, and nitrogen—they are non-directional, a fatal defect for the assembly of biomolecules with defined shapes. As Robert E. D. Clark noted, non-directionality is a fundamental reason why organisms could not use ionic bonds:

> The attraction between charged atoms is completely devoid of any *directive* quality: the atoms are only concerned to maintain the same *distance* between one another and, apart from this, they care nothing for organization.
>
> To make this point clear we may imagine a kind of Alice-in-Wonderland teapot, on the atomic scale. We will consider the teapot to be a "compound" of pot, spout, lid and handle. But when we put the lid, spout and handle on the pot we find, to our consternation, that they do not stay where we put them, but proceed to slip round the pot, beneath it and above!... In short, considered as an *organized* structure, the teapot fails to fulfill a useful function—its parts might just have well have been joined together by string!
>
> Fantastic as it sounds, the analogy gives us a passable picture of how atoms held together by electric charges [ionic bonds] behave.[23]

As Atkins points out, the distinction between these two types of strong bonds, ionic (non-directional) and covalent (directional), corresponds with the fundamental division between the inorganic domain and the organic domain, with the latter literally resting on the former:

> In general, molecular compounds [made up of covalently bonded atoms] are the soft face of nature, and ionic compounds [inorganic] are the hard. Few distinctions make this clearer than those between the soft face of the Earth—its rivers, its air, its grass, its forests, all of which

are molecular—and the harsh substructures of the landscape, which are largely ionic. This is why the upper triangle of the Eastern Rectangle [in the periodic table] is so important to the existence of life, and why all the rest of the kingdom is so important in the formation of a stable, solid platform.[24]

That the periodic table of elements should contain, in the region Atkins calls "the upper triangle of the Eastern Rectangle," a set of atoms including carbon (C), nitrogen (N), oxygen (O), and hydrogen (H), as well as phosphorus (P) and sulfur (S), possessing bonds of just the right strength for chemical manipulation in the cell as well as the crucial directional property, is surely indicative of a deep fitness in nature for carbon-based life.

And it is worth noting that these same atoms also form covalent directional bonds (more properly termed coordinate bonds) with a special class of metals—the transitional metals, including notably iron and copper. The directional bonding with the transitional metal atoms is also of vital importance, because it enables the metal atoms to be bound to proteins in unique molecular geometries, which in turn confers on the metal-protein complex many unique chemical properties that underlie specific enzymatic or other activities in the cell.

Weak Bonds

ALTHOUGH THE spatial arrangement of the atoms in the individual basic building blocks—amino acids and nucleotides, sugars, etc.—are determined by strong directional covalent bonds, the higher-order spatial deployment of the building blocks themselves (and their constituent atoms) in the major types of macromolecules is determined by much weaker chemical forces. These include van der Waals forces as well as the weak bonds, which involve electrostatic interactions between atoms and molecules, interactions which do not involve the sharing of electrons. The weak bonds are ten to twenty times weaker than strong or covalent bonds. Where the energy of a C-C covalent bond is about 350 kJ /mol1, the weak bonds have energies of 4–40 kJ/mol.[25]

Because of their role in determining the 3-D shape of macromolecules, the weak bonds are, as Watson describes them in *The Molecular Biology of the Gene*, "indispensable to cellular existence."[26] Their "indispensability" was only realized with the molecular biological revolution. Linus Pauling suspected that the weak bonds would prove pivotal, and his suspicions were confirmed by the discovery of the importance of macromolecular shape. As Crick explained in a speech honoring Pauling, chemists before World War II "were much more concerned with the strong bonds," and it was some time before they became interested in weak bonds. The weak bonds, as it turns out, "are the ones that fit molecules together in a physical sense, in one way or another, with different degrees, of course, of specificity. Pauling believed that that was going to be the key to everything, the way that things fitted together, using these weak forces."[27]

It is the weak bonds that fit together two different parts of an individual large macromolecule (like a protein), or two different molecules (as in the case of the two strands of the double helix), conferring unique spatial architectures on the resulting macromolecular complexes.[28] Watson nicely captured the nature and biological function of the weak bonds when he described them as conferring "selective stickiness"[29] to bio-matter, a stickiness that determines the way complex molecular structures fit together.

The double helical structure of DNA illustrates the way the strong directional and weak bonds work together to determine the overall atomic architecture of complex macromolecules. In the helix it is the strong directional bonds which determine the spatial position of the atoms in each of the nucleotides in each strand, while it is the job of the weak bonds to hold the two stands of DNA together in the classic, higher-order, 3-D double helical conformation. Consequently, the spatial positioning of the atoms in the double helix is determined by these two very different types of bonds working together.

Reversibility

THE DOUBLE helix also illustrates a major and vital characteristic of the weak bonds—that they can be broken relatively easily and are readily reversible. This characteristic enables the cell to pull apart the two strands of the helix during DNA replication and transcription, and enables them to easily reattach afterwards. If the bonds linking the two strands together were strong covalent bonds, the two strands would be irreversibly bound together; replication and transcription would be practically impossible.

Most biological functions not only require stereospecific arrangement of atoms in macromolecules but also (as with DNA) reversible weak binding between various molecular surfaces. And it is the relative weakness of the weak bonds which makes this possible.

Selective bonding between two molecular surfaces can only be achieved by using a number of bonds which collectively form a unique, complementary lock-and-key pattern of electrostatic interactions linking the two surfaces together. While it would be possible to bind two complementary molecular shapes or surfaces together with several strong covalent bonds, it would be difficult to pull the two molecules apart once the strong bonds had been made. That is, it would be difficult to remove the key from the lock. And even if the energetic barrier could be overcome, there would be the additional steric problem of fitting some "bond breaking" molecular device between the lock and the key to break the individual covalent bonds.

If the bonding between the two complementary molecular surfaces is to be both selective (comprising a number of bonds arranged in a unique pattern) and reversible (essential to most cellular functions), then the individual bonds must not be too strong or their collective action would bind the two surfaces into a rigid, immobile structure. Hence selective, reversible bonding could not be achieved using strong covalent bonds. The only way to achieve both highly specific and weak, reversible lock-and-key bonding between two molecules is to use the combined action of several much weaker bonds.

In short, the relative weakness of the weak bonds (compared with the strong bonds) is exactly what is needed for rapid, highly selective association and disassociation of two complementary molecular surfaces—between the two strands of DNA, between two stretches of the polypeptide chain of a protein, between an enzyme and its substrate, etc.

If weak bonds had been, say, ten times stronger, close to the energy of covalent bonds, selective stickiness would remain possible, but it would be irreversible.[30] Biochemistry as it occurs in cells would be impossible. Proteins and all the constituents of the cell would be frozen into rigid, immobile structures, including the two strands of the double helix.[31]

As Watson noted in *Molecular Biology of the Gene*, the strength of the weak bonds is "not so large that rigid lattice arrangements develop within a cell—the interior of a cell never crystallizes as it would if the energy of... [weak bonds] were several times greater." This fact "explains why enzymes can function so quickly, sometimes as often as 10^6 times per second. If enzymes were bound to their substrates by more powerful bonds, they would act much more slowly."[32]

On the other hand, if the strength of weak bonds were lower, then it would be impossible to fit enough of them on complementary surfaces to form a strong enough bond to withstand the hurly-burly and continual thermal jiggling of the constituents within the cell. Reversible, specific binding of a substrate to its catalytic site could never occur, nor, for example, could the head of a molecular motor detach from and reattach so readily to an actin fiber. Rob Phillips points out that as things are, disruptive forces in the cell caused by the random collisions between particles are close to the forces exerted by the weak bonds, meaning the weak bonds couldn't get much weaker and still do their job.[33]

In short, for reversibly sticking molecules together in the cell in highly specific stereospecific complexes—the vital basis of virtually all biochemical functions—the average energy level of the weak bonds has to be very close to what it is.

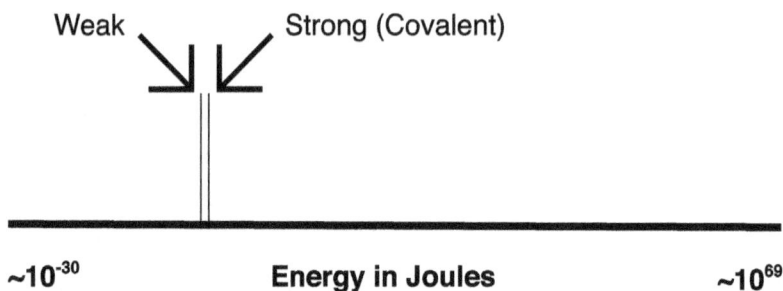

Figure 3.3. Energy levels in the cosmos vary over a vast range of magnitudes, from low-frequency photons ($\sim 2 \times 10^{-30}$ J) to the Big Bang ($\sim 4 \times 10^{69}$ J). The energy of one hydrogen bond is about 4–40 kJ/mol $= 6.7$–67×10^{-21} J, and the average energy of the strong bonds is about ten times greater (one order of magnitude more).

Also, the absolute strength of the weak bonds and their strength relative to that of the strong bonds must be very close to what they are.

Some idea of the exacting nature of this example of fine tuning can be appreciated by plotting energy levels in the cosmos on a logarithmic scale from the energy of low-frequency radio photons to the energy of the explosion which gave rise to the expanding universe, the "big bang"—a range exceeding one hundred orders of magnitude. In doing so, we find that on our graph[34] we can only represent the energy range of the strong and weak bonds as two infinitesimally narrow lines.

Materialism Undermined

OVER THE past two centuries there were two major discoveries which undermined previous vitalistic notions. The first was the discovery that the vital properties of organic compounds resided not in vital agency but in the unique chemical fitness of the carbon atom—along with oxygen, nitrogen, and hydrogen—to generate a vast inventory of compounds of diverse chemical and physical properties. The second was the discovery that the seemingly mysterious vital abilities of cells to carry out enzymic catalysis and to replicate genetic information largely depended on the fitness of two types of chemical bonds which uniquely enable the precise deployment of atoms into extraordinarily complex highly specific 3-D conformations.

For many people, the retreat of vitalism in biology in the two centuries following Wöhler has been a grand victory for positivistic thinking and materialism. Watson waxes lyrical in this vein in his book *DNA: The Secret of Life*. The discovery of the double helix was so important because it "brought the Enlightenment's revolution in materialistic thinking into the cell," he writes. "The intellectual journey that had begun with Copernicus displacing humans from the center of the universe and continued with Darwin's insistence that humans are merely modified monkeys had finally focused in on the very essence of life. And there was nothing special about it. The double helix is an elegant structure, but its message is downright prosaic: life is simply a matter of chemistry."[35]

Philosopher Daniel Dennett echoed these sentiments when he crowed, "Vitalism—the insistence that there is some big, mysterious extra ingredient in all living things—turns out to have been not a deep insight but a failure of imagination."[36]

But Watson, Dennett, and their fellow materialists should be more circumspect. When they celebrate each retreat of vitalism as a great victory for materialism and mechanism, they overlook a crucial point: each retreat revealed some additional element of intelligent fine tuning in nature—fine tuning for life.

Yes, vitalism retreated, but in its place the wonder of the unique capabilities of the carbon atom and its peerless fitness for biochemistry was revealed. Yes, vitalism retreated with the discoveries which followed in the wake of the molecular biological revolution, but these only highlighted a greater wonder in the fitness of nature for the assembly of complex macromolecules. These were not wonders beyond scientific analysis, but they were wonders nonetheless, and they pointed not to a homunculus in the cell, but to a far greater wonder-worker who finely tuned the very fabric of nature for life on Earth.

4. Carbon's
Collaborators

Just as the letters of the alphabet have the potential for endless surprise and enchantment, so, too, do the elements of the kingdom. Unlike an alphabet which has hardly any infrastructure, the kingdom has sufficient structure to make it an intellectually satisfying aggregation of entities. And because these entities are finely balanced, living personalities, with quirks of character and not always evident dispositions, the kingdom will always be a land of infinite delight.

—Peter Atkins, *The Periodic Kingdom*[1]

THE PREVIOUS TWO CHAPTERS HAVE DESCRIBED MANY ELEMENTS of fitness of the carbon atom for life. But the recipe for life requires more than the carbon atom. Carbon needs collaborators to build the great plenitude of compounds to assemble a complex living system. And, as usual, nature has obliged.

Although most of the atoms in the periodic table are metals and do not form strong covalent directional bonds, carbon's nonmetal near neighbors in the table, oxygen (O) and nitrogen (N), clustered together near the top right-hand corner, and hydrogen (H), at the upper left, share carbon's ability to form strong directional covalent bonds (see Figure 4.1). In one of the classic papers arguing for the fitness of nature for carbon-based life, George Wald (awarded a Nobel prize for elucidating the molecular basis of photodetection) emphasized their special fitness for biochemistry:

The special distinction of hydrogen, oxygen, nitrogen, and carbon is that they are the four smallest elements in the Periodic System that achieve stable electronic configurations by gaining, respectively, 1, 2, 3, and 4 electrons. Gaining electrons, in the form of sharing them with other atoms, is the means of making chemical bonds, and so of making molecules. The special point of smallness is that these smallest elements make the tightest bonds and so the most stable molecules; and that carbon, nitrogen, and oxygen are the only elements that regularly form double and triple bonds. Both properties are critically important.[2]

Not only do hydrogen, oxygen, and nitrogen share with carbon the prime characteristics of fitness for building stable molecules of defined 3-D shape, but their chemical and physical properties are also markedly different from carbon. This diversity is crucial because it allows these elements to introduce into the organic realm carboxyl (COOH), amino (NH_2), methyl (CH_3) and other groups with novel chemical properties.

If carbon's nonmetal neighbors in the second row (period 2) of the periodic table, nitrogen and oxygen, had possessed physical and chemical properties similar to carbon (as with most adjacent atoms in most regions of the periodic table), the organic realm would have been much less chemically diverse. Carbon-based life may even have been impossible, or at least restricted to simple unicellular life. But nitrogen, carbon's near

↓ Period																	
1	1 H																	2 He
2	3 Li	4 Be											5 B	6 C	7 N	8 O	9 F	10 Ne
3	11 Na	12 Mg											13 Al	14 Si	15 P	16 S	17 Cl	18 Ar
4	19 K	20 Ca	21 Sc	22 Ti	23 V	24 Cr	25 Mn	26 Fe	27 Co	28 Ni	29 Cu	30 Zn	31 Ga	32 Ge	33 As	34 Se	35 Br	36 Kr
5	37 Rb	38 Sr	39 Y	40 Zr	41 Nb	42 Mo	43 Tc	44 Ru	45 Rh	46 Pd	47 Ag	48 Cd	49 In	50 Sn	51 Sb	52 Te	53 I	54 Xe
6	55 Cs	56 Ba		72 Hf	73 Ta	74 W	75 Re	76 Os	77 Ir	78 Pt	79 Au	80 Hg	81 Tl	82 Pb	83 Bi	84 Po	85 At	86 Rn
7	87 Fr	88 Ra		104 Rf	105 Db	106 Sg	107 Bh	108 Hs	109 Mt	110 Ds	111 Rg	112 Cn	113 Nh	114 Fl	115 Mc	116 Lv	117 Ts	118 Og

Lanthanides	57 La	58 Ce	59 Pr	60 Nd	61 Pm	62 Sm	63 Eu	64 Gd	65 Tb	66 Dy	67 Ho	68 Er	69 Tm	70 Yb	71 Lu
Actinides	89 Ac	90 Th	91 Pa	92 U	93 Np	94 Pu	95 Am	96 Cm	97 Bk	98 Cf	99 Es	100 Fm	101 Md	102 No	103 Lr

Fig 4.1. The Periodic Table of the Elements.

neighbor to the right, and oxygen, the neighbor of nitrogen a step further to the right, are about as different from carbon as can be imagined, being transparent, colorless gases in ambient conditions—very different from a pile of soot or a lump of coal, two common forms of carbon in nature. (Hydrogen, carbon's other partner, is also a colorless gas in ambient conditions.)

Also important for life is how different carbon and its period 2 partners are from neighbors outside period 2. Lawrence Henderson stressed this point, noting that these period 2 elements "possess very definite individual properties, which mark them off sharply from other substances." It's thus "in the highest degree probable that compounds made from elements of such positive chemical characteristics and very unusual properties will be unlike compounds formed from other elementary substances." Henderson concludes that this leads "us to believe that other elements are exceedingly unlikely to readily form compounds comparable in number, variety, and complexity with those of organic chemistry as we know it."[3] He further elaborates: "It follows from the peculiarities just explained that the first great factor in the complexity of living organisms as we know them, the complexity and variety of their chemical constituents, depends principally upon the nature of the elements which compose such substances, and is most probably a unique, certainly a very rare characteristic of matter."[4]

In a similar vein, Peter Atkins emphasizes that all the neighboring nonmetals atoms in period 2—boron, carbon, nitrogen, oxygen, and fluorine—differ markedly from their homologues immediately below in period 3. Nitrogen, he notes, "is a colorless nonreactive gas," while its nearest neighbor in the third period "is phosphorus, a colorful reactive solid, while sulfur, directly below the colorless gas oxygen, "is a yellow solid."[5] Indeed, as he further notes, all the atoms of period 2, including boron, carbon, and fluorine are strikingly different from their immediate southern neighbors.[6]

Only among carbon's near neighbors in the small, nonmetal region of the periodic table are there atoms able to make strong directional covalent bonds and that possess very different properties, capable of making the right stuff for life when combined with carbon. In this region, *the chosen region*, as Atkins puts it, "Complexity can effloresce from subtly different consanguinity."[7]

In sum, carbon's genius is to a large degree gifted by its covalent, nonmetal partners, nitrogen, oxygen, and hydrogen, and their differing properties. In fact, it turns out that—consistent with the special fitness of nature for life on Earth—the properties of carbon's collaborators are precisely what are needed for the chemistry of the cell.[8]

Electronegativity

THE HIGHER an atom's electronegativity, the greater its attraction for electrons. Of all the similarities and differences among the chemistry of hydrogen, carbon, nitrogen, and oxygen, one of the most consequential involves their electronegativities.

Consider first the carbon-hydrogen bond[9] and the chemical consequences of the fact that these two atoms have similar electronegativities.

As a consequence, when hydrogen atoms bond to carbon (C-H), they form what are called non-polar covalent bonds. In these bonds, the electrons are equally distributed between the two atoms. And because of this, there is little or no charge disequilibrium in the resultant bond. In other words, the bonds are electrically symmetrical. There are no electropositive regions (with a deficit of electrons) and no electronegative regions (with a surplus of electrons) around the bond. In C-H bonds, the constituent atoms tug more or less equally on their electrons, the charge distribution remains uniform, and the resulting compounds are non-polar thanks to sharing their electrons equally. As we will see below, one of the effects of this is to render hydrocarbon chains insoluble and hydrophobic, a characteristic which plays a vital role in the life of the cell.

And now we come to something very beautiful, and of the profoundest consequence to all life on Earth. Unlike the bonds that hydrogen makes with carbon, which are non-polar,[10] the bonds that hydrogen (H) makes with nitrogen (N) and oxygen (O) are nearly always polar, involving unequal sharing of electrons. This is because, unlike hydrogen and carbon—which are relatively close on the Pauling electronegativity scale (0.35 units apart)—the electronegativity of hydrogen differs markedly from that of nitrogen or oxygen (0.7 and 1.3 units apart respectively). Because of the unequal attraction of the electrons, the oxygen and nitrogen atoms in O-H and N-H bonds are negatively charged while the hydrogens are positively charged.

One of the most important of polar molecules is water (H-O-H), the matrix of life. The polar nature of water molecules is responsible for its great power as a solvent in dissolving charged ions or polar compounds. When water molecules come in contact with a charged ion or polar compound, the water molecules experience electrostatic interactions (charge-based attractions). The negatively charged oxygen atoms are attracted to positively charged ions such as Na^+ or regions of electrostatic positivity in a polar compound, while the positively charged hydrogen atoms are attracted to negatively charged ions such as chloride Cl^- or regions of electronegativity in polar compounds. We call such ions and polar molecules *hydrophilic* ("water-loving").[11]

As there are nearly always many water molecules relative to solute molecules, these interactions lead to a sphere of water molecules around the solute. These hydration shells allow particles to be dispersed evenly in water and keep the charged solutes apart and unable to combine and precipitate out of the solution and, therefore, soluble. As the majority of organic molecules—sugars, alcohols, fatty acids, amino acids, nucleotide bases—contain charged groups or are polar, they lend themselves to the formation of hydration shells and are readily soluble in water. And as water is the matrix of the cell, the solubility of so many compounds is of great utility.

But what happens when non-polar, long-chain hydrocarbons try to join the electrostatic party? Because of the absence of charged regions, water molecules cannot form hydration shells around the molecules, and the hydrocarbon chains are excluded from the party and forced into insoluble, water-avoiding clusters. For this reason, long-chain hydrocarbons are termed *hydrophobic* ("water-fearing"). And the force which clumps them together into water-excluded clusters is termed the *hydrophobic force*.

But rather than being a defect in the order of things, the insolubility and hydrophobic character of hydrocarbons is a vital element of nature's fitness for life. The insoluble hydrocarbons form the core of the lipid bilayer membrane that surrounds the cell and many of the cell's internal organelles—one of the most important supramolecular structures in the cell, indeed in the entire biological world. It also allows newly synthesized proteins to fold into their secondary structures, one of the most important of all the biochemical processes in the cell.

The basic building blocks of the cell membrane consist of long, insoluble, hydrophobic hydrocarbon chains linked to a polar hydrophilic group or head containing O-H and N-H bonds. Because they contain a hydrophobic group linked to a hydrophilic group, they are termed *amphiphilic*, from the root words *amphi* (meaning "both") and *phileo* (meaning "to love"). These building blocks are known as phospholipids, because they contain a phosphate group in the hydrophilic head. Because of their hydrophobic character, the hydrocarbon chains or tails are forced to cluster together into two layers in the center of the membrane (away from water), with the water-loving hydrophilic groups clustered at the interface between the hydrocarbon bilayer and the aqueous medium on the inside and outside of the cell.

As John Philip Trinkaus comments in his wonderful book *Cells into Organs*, "Because water is itself a strongly polar molecule, the polar phosphate of the membrane lipids will inevitably be attracted to the surfaces of the membrane, both external and cytoplasmic. And just as inevitably their nonpolar fatty acid parts will tend to be squeezed into a nonpolar phase in the interior of the membrane."[12] It is this dual character of the phospholipids that enable them to form membranes, without which cells would not exist.

The insoluble hydrophobic hydrocarbon's other important role in the cell is in protein folding. Many amino acids possess a hydrophobic,

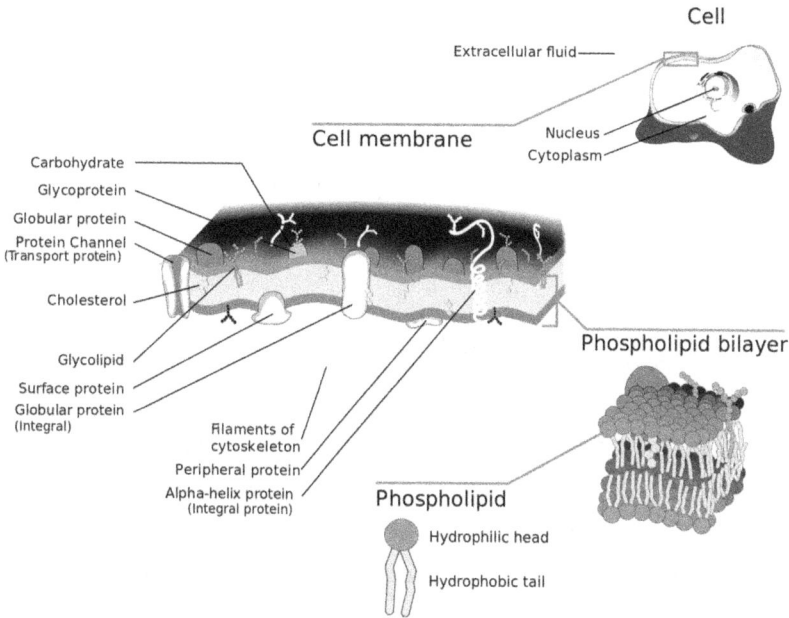

Figure 4.2. The lipid bilayer cell membrane.

insoluble, hydrocarbon side chain (leucine and isoleucine are two examples). However, all amino acids (including those with hydrophobic side chains) also contain the two polar groups COOH and NH_2. This means that the hydrophobic amino acids, like the phospholipids in the membrane, are also amphiphilic and soluble to a degree. The same hydrophobic force that causes the hydrophobic tails of the phospholipids to cluster together in the center of the membrane forces the hydrophobic non-polar side chains to cluster into a central core, playing a decisive role in protein folding.

In an additional teleological twist, these water-avoiding aggregates that are clustered together in the center of the folded protein provide the cell with tiny, non-aqueous micro-environments vital to the life of the cell. Many of the synthetic and enzymatic processes on which the cell depends can only occur in a chemical micro-environment free of water. So, the relative electronegativities of C, H, and O are responsible for the hydrophobic force, which plays a vital role in the folding and function of proteins.

Charles Tanford was not exaggerating when he said, "The hydrophobic force is the *energetically* dominant force for containment, adhesion, etc., in all life processes," adding: "This means that the *entire* nature of life as we know it is a slave to the hydrogen-bonded structure of liquid water."[13]

A Finely Tuned Quartet

THE WAY the different electronegativities of hydrogen, carbon, oxygen, and nitrogen work together towards the formation of the cell membrane and the folding of proteins is amazing. On the one hand, the electrical asymmetry of oxygen-hydrogen bonds leads to the hydrophilic character of water and is the source of the hydrophobic force, which clumps the insoluble non-polar hydrocarbons into the bilayer membranes and clumps the hydrophobic amino acid side chains into the center of proteins. On the other hand, the electrical symmetry of carbon-hydrogen bonds makes the clumping possible by conferring on long hydrocarbon chains their non-polar, hydrophobic, water-avoiding behavior.

This is so elegant. In the face of the wondrous way nature works such a miracle, one can only cite John Keats's Ode *On a Grecian Urn*: "Beauty is truth, truth beauty."[14]

At the heart of cellular life is an extraordinary reciprocal fitness between the non-polar carbon-hydrogen (C-H) bonds and the polar oxygen-hydrogen (O-H) bonds. This reciprocity gifts life with the cell membrane and the folding of proteins. If the electronegativity of hydrogen, carbon, oxygen, and nitrogen had been the same, unquestionably there would be no carbon-based life on Earth. The cosmos has only come to life because of the different electronegativities of the four collaborators.

And as we have seen, in another teleological twist, it is hydrogen's electronegativity that is the cause of water's polar nature (via the electrical asymmetry of the O-H bond) and of the non-polar nature of the long hydrocarbon chains (because of the electrical symmetry of the C-H bond). Hydrogen's place in Linus Pauling's electronegativity scale is responsible for both the hydrophobic character of long hydrocarbon chains and for the character of water that renders them insoluble.

Carbon-based life likely would be impossible if the electronegativities of hydrogen (H), carbon (C), oxygen (O), and nitrogen (N) were even slightly different from what they are. So, for example, imagine a world where the electronegativities of these four elements were closer to one another, a world devoid of polar molecules. Alternately, envision a world where C-H bonds were polar and O-H and O-N bonds non-polar. Neither of these imagined worlds would contain carbon-based life, even if all the other properties of these four elements were exactly the same. And not because we lack the imagination to see how life could manage in these counterfactual worlds. Just the opposite. We can assert the impossibility precisely because molecular biologists have done the painstaking work of uncovering the wonder manifest in this unique band of atoms.

The unique capacity of carbon to bond with itself, its capacity to form multiple bonds, the metastability of so many carbon compounds, the directionality and strength of the covalent bonds of carbon and its nonmetal compatriots, the existence of weak chemical forces such as van der Waals forces and weak ionic bonds of appropriate strength for lock-and-key bonding—all these would be useless without the fine-tuning of the relative electronegativities of oxygen, nitrogen, and hydrogen. Only when the whole suite of fitness is complete can the miracle of the cell be actualized.

And as for water's inability to solubilize oils and fats and other hydrocarbons, this might seem a defect in the so-called *universal solvent*. But as we have seen, the hydrophobicity of hydrocarbons, and their insolubility in water, is one of the prime elements of nature's fitness for carbon-based life. An apparent shortcoming of water turns out to provide a previously unsuspected but vital element of nature's fitness for carbon-based life.

The Fitness of the Cell Membrane: A Checklist

The membrane's many particular properties that make it so fit to form the bounding membrane of the cell are worth looking at individually.

Semipermeability: Perhaps the most important function of the membrane is providing the cell with a semipermeable bounding membrane to separate it from the external environment.

No cell could survive without some sort of membrane that was relatively impermeable to the cell's constituents, especially small metabolites such as sugars and amino acids but also larger molecules such as proteins and RNAs. The cell needs a way to prevent these from diffusing away into the surrounding fluid. It must have been this way from the beginning. The very first cells must have been enclosed by a membrane to retain the mix of biochemicals within.

The membrane is non-polar, which makes it particularly impervious to polar and charged compounds. This is vital because the small organic molecules in the cell—a mix of sugars, nucleotide bases, metabolic acids, etc.—are nearly all polar molecules. So the lipid bilayer is fit to provide not only a barrier between the inside and outside of the cell, but one that is non-polar—precisely the sort needed to retain the small charged metabolites inside the cell.

At the same time, the membrane isn't perfectly impermeable. That would make cell survival equally impossible, preventing it from getting nutrients and expelling waste. This problem is solved by the existence in the membrane of special gates or pores through which myriads of small compounds and ions pass in and out of the cell in a highly controlled manner (see below).

Self-Organizing: A second emergent property of the membrane is its self-organizing ability. This remarkable property allows the membrane to form automatically around the outer surface of the cell. As Trinkaus comments in *Cells into Organs*, "Because of their intrinsic chemical nature... phospholipids *naturally* and *spontaneously* self-assemble... to form a bilayer in a watery solution, sequestering their hydrophobic ends in the interior while their polar hydrophobic ends are 'in solution' on the surface. It is as it were 'the nature of the beast' for them to do so."[15]

In addition, any bounding membrane must possess the ability to self-repair so it can readily close any hole or break that would allow the cell's contents to diffuse away. The membrane needs to be able to adapt to the minute-to-minute deformations in cell form and repair itself to maintain a continuous barrier at all times in the face of all sorts of contingencies. It's not hard to imagine the huge problems a cell would face if its bounding membrane were not self-mending and did not quickly close any hole or defect in its surface.

The lipid bilayer wonderfully satisfies the necessity for both self-assembly and self-repair. As Trinkaus explains, it can flow in every direction over the cytoplasmic surface so as to maintain a continuous barrier between cell and surroundings, and it manages this in the face "of the ever-changing protrusive activities of the cell surface."[16]

Trinkaus describes the membrane as a "two-dimensional liquid"[17] and comments, "It is of great significance for cell motile behavior that the lipid bilayer of the plasma membrane has an overall high fluidity (low viscosity) at the normal body temperature of homeotherms."[18]

It is therefore not incidental that every cell on Earth has a lipid bilayer forming the skin or boundary layer. Its qualities of impermeability (except at the gates), relatively high fluidity, and spontaneous assembly and self-repair appear to be essential to a cell's existence. This unique combination of characteristics is only found in the lipid bilayer, another case where a crucial biological function, the bounding of the cell, is carried out by a structure that appears to be both unique and ideal for its assigned role. No other known material could substitute for this particular structure.

The cell would encounter profound challenges if it used a membrane that was not self-organizing or had to be assembled piece by piece. Not only would assembly be a colossal logistical and energy-demanding challenge; the cell also would need a complex sensing system to inform it of the minute-to-minute changes in membrane conformation. Clearly, a mechanical skin that could not self-organize would never work as a barrier surrounding the ever-shifting shape of a living cell.

While crawling, a cell continually changes shape, especially at the leading edge. This is known as the lamellipodium, which is full of actin fibers that grow to propel the cell forward. During crawling, many cells create micro-protrusions or filipods—narrow cylindrical extensions containing actin fibers—which they use like a cat's whiskers to detect environmental clues, initiate contact with surrounding cells, and convey signals back to the cell body.[19] [20]

Filipods are used, for example, during axon growth and guidance by path-finder neurons to find their way in the developing nervous system.[21] Plant, fungal, and most bacterial cells, encased in rigid cell walls, have less need for the self-sealing, self-organizing properties of the cell membrane. But in cases where cells undergo continual changes in form, as when an amoeboid white cell chases down a bacterium in the bloodstream or during cell movements in embryogenesis, the self-sealing, self-organizing ability of the membrane is essential.

Without the innate self-organizing property of lipid bilayer membranes, there would be no crawling, no embryogenesis, and probably no thinking beings to contemplate the loss.

Membrane Diversity: The self-organizing nature of the membrane also facilitates the generation of a great variety of emergent structures, including tubes, vesicles, and various types of layered structures.[22]

The different morphologies of the membranes in the chloroplast and those constituting photoreceptive discs in the outer segments of the photoreceptors in the vertebrate eye testify to the adaptive versatility of this remarkable structure. The same basic membrane structure makes up the endoplasmic reticulum, encloses the nucleus and the mitochondria, and so forth. As I wrote in *Biology and Philosophy*:

> All this plethora of forms arises spontaneously by self-organization from the physical properties of various bilayer lipid membranes of differing chemical composition. Different lipids and proteins can bend and distort the basic membrane form into various globular forms, vesicular

forms or tubules. The various forms arise again like the protein folds entirely from the local interactions between membrane constituents. A zoo of vesicular forms can be generated spontaneously from lipid membranes depending on the ratio of lipid and protein constituents. This is the focus of a lot of current work. As Huttner and Schmidt comment: "The shape of biological membranes reflects the shape of its principle constituents—that is membrane lipids and integral membrane proteins, as well as their interaction with each other and with peripherally associated proteins (including the glycoproteins and proteoglyans), the cytoskeleton and the extracellular matrix."[23]

And in reporting some fascinating work on a membrane-altering protein dynamin, Wieland Huttner and Anne Schmidt remark, "Dynamin alone is sufficient to change the shape of liposomes, causing either tubulation or vesiculation, depending on lipid composition."[24] Thus, by altering the protein and lipid constituents of membranes, the cell can generate vesicles from planar surfaces, tubules from planar surfaces, and vesicles from tubules. The intrinsic properties of cell membranes generate a vast zoo of emergent membrane structures.

The Right Width: Another crucial property of the lipid bilayer is how remarkably thin it is, only five nanometers.[25] If a typical mammalian cell were expanded to the size of a pumpkin, its lipid bilayer membrane would be thinner than a sheet of paper. But despite its thinness, it is highly robust and can maintain its oily integrity in the face of the varied micromechanical forces buffeting it.

The remarkable thinness of the membrane isn't just a striking curiosity. It's essential to the cell's fitness. Proteins average about five nanometers across, about the same as the width of the membrane. And this means that individual proteins can be designed to stretch across the thickness of the membrane. This is important because there are several membrane functions which require individual proteins which can span the width of the membrane. Some make up the ion channels that selectively control the passage of ions—mainly Na^+ and K^+—across the membrane. Others maintain different ion concentrations on the two sides of the membrane. They also couple the transport of ions with that of small organic molecules,[26] relaying signals between a cell's interior and exterior. Other important transmembrane proteins are those involved in cell-to-cell recognition and selective adhesion.[27]

Several intriguing elements of fitness determine the width of the membrane. For example, the membrane requires pliable, relatively stable hydrocarbon chains. The chains also need to be sparingly soluble. The chain length of the hydrocarbons used in membranes in the cell is between sixteen and eighteen carbon atoms long. Hydrocarbon chains of more than eighteen carbon atoms long are at ambient temperatures too stiff, insoluble, and wax-like. They cannot be readily mobilized in water and are not sufficiently fluid or pliable for the membrane. On the other hand, chains less than sixteen carbon atoms long are too mobile and unstable. But chains of sixteen to eighteen are just right.[28] As Arthur Needham points out, one factor restricts chain-length to some degree, while additional factors further restrict the functional range:

> The chains are held in orientation by the London-van der Waals attractions of neighbouring molecules and this is effective only within a certain range of chain-lengths, namely C_{16} to C_{36} or so; this is precisely the range of the common biological fatty acids... Beyond this length the molecules topple and tangle, while shorter chains have insufficient attraction... If the lipid is also polar, like the fatty acids, that is if it has a hydrophil group in the molecule, then members shorter than C_{16} also tend to pass into aqueous solution too readily to form stable membranes.... The C_{18} fatty acids are much the most abundant in plants and animals... C–even members up to C_{38} have been discovered in biological material but they only serve to emphasise the unique exploitation of the C_{16-18} node. Anything shorter does not build stable crystalline layers and anything longer is less soluble in water, protects proteins less well, has too high a melting point, and so on.[29]

How fortunate it is that these various essential factors overlap in the sixteen-to-eighteen range of chain length. The two C_{18} hydrocarbon chain layers result in the five-nanometer width of the membrane, just the right width for individual proteins to stretch across.

Membrane Potential: Because the membrane—or more specifically the central hydrocarbon layer—is an electrical insulator, an electric charge can be built up across the membrane. The difference in electric charge across the cell membrane is known as the membrane potential. Cells generate the potential by regulating the movement of charged ions across the membrane through highly selective ion channels.

Resting mammalian cells may use 30 percent of their metabolic energy in generating the membrane potential, which may rise up to 70 percent[30] in nerve cells. This may seem like a prodigal use of cellular energy, but the charge serves many important functions, as Crichton explains:

> Many transmembrane transporter proteins termed secondary transporters, use the discharge of an ionic gradient to power the "uphill" translocation of a solute molecule across membranes. Coupling solute movement to ion transport enables these secondary transporters to concentrate solutes by a factor of 10^6 with a solute flux 10^5 faster than by simple diffusion... Sugars and amino acids can be transported into cells by Na^+-dependent symports. Dietary glucose is concentrated in the epithelial cells of the small intestine by a Na^+-dependent symport, and is then transported out of the cells into the circulation by a passive glucose uniport situated on the capillary side of the cell.[31]

The cell's capacity to generate and maintain a charge across its membrane allows for the transmission of nerve impulses along the axons of neurons in animal nervous systems. The transmission of a nerve impulse depends on what is called the action potential. This is an extremely rapid depolarization of the membrane, which occurs via a sudden influx of millions of Na^+ ions into the cell in a fraction of a millisecond.[32] The impulse or action potential travels along the axon at speeds up to one hundred meters a second. As Bruce Alberts and his co-authors explain:

> In nerve and skeletal muscle cells... Na^+ channels... open, allowing a small amount of Na^+ to enter the cell down its electrochemical gradient. The influx of positive charge depolarizes the membrane further, thereby opening more Na^+ channels, which admit more Na^+ ions, causing still further depolarization. This process continues in a self-amplifying fashion until, within a fraction of a millisecond, the elec-

trical potential in the local region of membrane has shifted from its resting value of about -70 mV to almost as far as the Na^+ equilibrium potential of about $+50$ mV.[33]

The membrane's electrical insulating character and the resulting membrane potential provides precisely the electrical characteristics required for the transmission of electrical impulses between cells and ultimately for the construction of the nervous system of beings like ourselves. This suggests, as in many other instances, that nature's fitness is not just for the carbon-based cell but also for advanced multicellular organisms like ourselves.

The generation of membrane potential and the nerve impulse depend critically on the mobility of small ions, which can move rapidly down concentration and electrical gradients. Their mobility and speed moving through ion channels is astounding. Alberts and his colleagues explain that "up to one million ions can pass through one open ion channel each second."[34] Without these tiny, highly mobile inorganic ions, no cell would be able to regulate or generate a membrane potential or a nerve impulse. No other small particles of matter possess charge and such great mobility. Neither proteins nor any of the organic molecules in the cell have the right properties to stand in for the alkali metal ions.

For the first time in this book, we have described a cellular function that involves atoms other than the four partner atoms carbon, hydrogen, oxygen, and nitrogen. Membrane potential and nerve transmission depend on the existence of billions of charged particles in the cell and the surrounding extracellular fluids, and on the very high diffusion rates of sodium and potassium ions.

The Quartet in Review

We have seen in the previous two chapters that the atoms carbon, hydrogen, oxygen, and nitrogen are impressively fit for the assembly of the vast plenitude of organic compounds with their diverse properties upon which the life of the cell depends. Among other things we saw that because of their directional strong covalent bonds, they form compounds with specific molecular shapes and that the existence of compounds with defined molecular shapes together with the weak chemical bonds enables the formation of large macromolecules of defined 3-D shape, macromolecules that can perform specific biochemical and cellular functions.

And as we have seen in this chapter, these four atoms possess just the right electronegativities to make polar molecules via the O-H and N-H bonds. They also provide non-polar bonds via C-H bonds, which render long-chain hydrocarbons hydrophobic and insoluble in polar fluids such as water, and this leads in turn to the emergence of the self-organizing wonder of the bilayer lipid cell membrane, which as we have seen has a whole suite of equally extraordinary properties, all of which are indispensable to the cell.

In short, these four atoms are mutually fit, in a myriad of improbable ways, to set in motion a succession of unique emergent phenomena, along a pathway stretching from the relative electronegativities of carbon, hydrogen, oxygen, and nitrogen to the nervous system of higher organisms. The way these elements work together to generate this unique path is stunning, conveying an irresistible impression of contrivance.

5. Energy for Cells

One salient characteristic of life is that it is not over in a flash but involves the slow uncoiling and careful disposition of energy: a spot of energy here, another there, not a sudden deluge. Life is a controlled unwinding of energy. Phosphorus, in the form of adenosine triphosphate (ATP) turns out to be a perfect vector for the subtle deployment of energy, and it is common to all living cells.

—Peter Atkins, *The Periodic Kingdom*[1]

Lake Mono in California is one of oldest lakes in North America, estimated to have first formed some 800,000 years ago. About eighteen kilometers long and fifteen wide, it sits in a dry desert basin nearly 2,000 meters above sea level between Yosemite National Park and the Nevada border. Mark Twain described its surroundings as a "lifeless, treeless, hideous desert."[2] Because it has no outlet to the sea, the dissolved salts in the runoff from the surrounding mountains remain in the lake. Over time, this process has raised its salinity to twice that of seawater and its alkalinity to that of a dilute solution of caustic soda. Hence its official designation as a "soda lake." Few organisms thrive in its caustic, hostile waters except for various species of algae adapted to high salinity, which provide food for the celebrated brine shrimp and the remarkable alkali flies which breed in the billions along the shore. But it is an even stranger species in this soda lake that would serve as the locus for a discovery with potentially profound bearing on the unique fitness paradigm.

Figure 5.1. Lake Mono, California.

Along with the algae, brine shrimps, and alkali flies, Lake Mono also contains an extremophile bacterium referred to by the acronym GFAJ-1. It was isolated from sediments on the lake bottom by NASA astrobiology fellow Felisa Wolfe-Simon, who named the organism after the first letters in the phrase "Give Felisa a Job."[3] What particularly intrigued Wolfe-Simon about GFAJ-1 and the extreme environment of Lake Mono is that, in addition to being hypersaline and alkaline, it boasts one of highest concentrations of arsenic in the world.

Subsequent studies of this new bacterium in Wolfe-Simon's lab raised, for the first time, the exciting possibility that GFAJ-1 might build its DNA and other biomolecules using arsenic as an alternative to phosphorous—the first instance of an organism using something other than the canonical six atoms of carbon (C), hydrogen (H), oxygen (O), nitrogen (N), sulfur (S), and phosphorus (P) to assemble its polymers and biomolecules. In a hastily arranged press conference in December 2010, organized by NASA, Wolfe-Simon made the dramatic announcement: GFAJ-1 could substitute arsenic (A) in place of phosphorus in its DNA if grown in cultures lacking phosphate but rich in arsenic. In other words, instead of using phosphate radicals (PO_4) to link together the successive nucleotides in DNA, the arsenic-eating bacterium apparently used arsenates (AsO_4).

The discovery was described in *Science* as having "profound evolutionary and geochemical importance."[4] And certainly, if verified, it would have been the first instance of an atom other than the canonical six being used to construct the key polymers of life.

The blogosphere went crazy, and articles hyped up the result, with titles like "Arsenic-Eating Microbe May Redefine Chemistry of Life"[5] and "Arsenic-Loving Bacteria May Help in Hunt for Alien Life."[6]

Others were more skeptical. Writing in the journal *Nature*, science reporter Erika Check Hayden noted, "If GFAJ-1 is indeed utilizing arsenic as Wolfe-Simon and her co-authors suggest, [Steven] Benner writes, the result would 'set aside nearly a century of chemical data concerning arsenate and phosphate molecules.' Benner... criticizes the paper for not fully taking into account how much existing science would need to be rewritten to accommodate its extraordinary claim."[7]

As it turned out, the claim was later shown to be false.[8] No one believes in arsenic-eating bacteria any longer. GFAJ-1 was just an ordinary life form after all.

However, this saga did have one intriguing and unintended consequence. The claim that arsenic might substitute for phosphorus in biomolecules fueled a discussion of the relative merits of phosphorus over arsenic, highlighting what has been well known for many years: phosphates are vastly more fit than arsenates for living chemistry, particularly in an aqueous environment.[9] Phosphorus compounds are durable in water while arsenic compounds fall apart quickly. How much more quickly? Experiments have shown that the half-life of arsenic linkages in DNA is 0.06 seconds compared to thirty million years for phosphate linkages.[10]

The denouement of the arsenic saga drew widespread attention to the chemical stability and unique fitness of phosphates for bioenergetics. Its superior fitness for this role has everything to do with the celebrated universal carrier and donator of chemical energy in living things, the ubiquitous compound *adenosine triphosphate* (ATP), the compound which is quite literally the fuel of life.

Energy—Spelled ATP

ALL CELLS need energy. They need energy to carry out their varied functional repertoire, including many different enzymatic actions, synthesizing their complement of proteins, lipids and DNA, and pumping ions across the cell membrane. They need energy to move, crawl over the substratum, and transport materials inside the cell, with molecular motors carrying cargoes along microtubules or actin filaments. Most of these activities—from crawling to synthesizing proteins and other polymers such as DNA—can only proceed if chemical energy is put into the system.[11] Generating and using energy is thus basic to all of cell biology.

The amount required per day is enormous. "We use about 2 milliwatts of energy per gram—or some 130 watts for an average person weighing 65 kg, a bit more than the standard 100 watt light bulb," Nick Lane writes. "That may not sound like a lot, but per gram it is a factor of 10,000 more than the sun (only a tiny fraction of which, at any one moment, is undergoing nuclear fusion). Life is not much like a candle; more of a rocket launcher."[12]

All cells on Earth use the energy locked up in the high-energy chemical bonds in the molecule ATP, and in other energy-rich phosphate molecules, to fund their chemistry and other activities. Lane comments:

The energy 'currency' used by all living cells is a molecule called ATP... ATP works like a coin in a slot machine. It powers one turn on a machine that promptly shuts down again afterwards. In the case of ATP, the 'machine' is typically a protein. ATP powers a change from one stable state to another, like flipping the switch from up to down. In the case of the protein, the switch is from one stable conformation to another. To flip back again requires another ATP, just as you have to insert another coin in the slot machine to have a second go. Picture the cell as a giant amusement arcade, filled with protein machinery, all powered by ATP coins in this way. A single cell consumes around 10 *million* molecules of ATP every second! The number is breathtaking. There are about 40 trillion cells in the human body, giving a total turnover of ATP of around 60–100 kilograms per day—roughly our own body weight. In fact, we contain only about 60 grams of ATP, so we know that every molecule of ATP is recharged once or twice a minute.[13]

One hundred kilograms of ATP works out to about 10^{29} molecules. This is an almost impossibly large number to imagine. It is more molecules than there are stars in the observable universe and about a billion times the number of seconds in the four billion years since the formation of the Earth. And some organisms generate and use even more ATP per cell per second than the body cells in humans. As R. K. Suarez shows, the case of a bee feeding on nectar is particularly impressive:

> During "normal" hovering, a steady-state activity hummingbirds engage in for the purpose of foraging on floral nectar, flight muscle O_2 consumption rates of approximately 2 ml g^{-1} min^{-1} have been estimated, corresponding to rates of ATP turnover of close to 500 μmol g^{-1} min^{-1}. Worker honeybees, hovering at wingbeat frequencies of 250 Hz [250 wingbeats per second], are even more impressive as their flight muscles consume 6 ml O_2 g^{-1} min^{-1}. A single flying honeybee turns over (i.e. hydrolyzes and resynthesizes) 1.39×10^{15} molecules of ATP per wingbeat cycle![14]

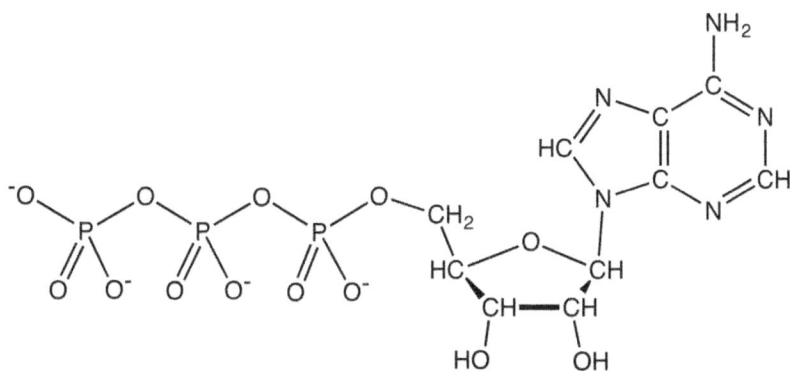

Figure 5.2. Adenosine triphosphate (ATP). The capacity of ATP to play its role as the energy currency of the cell is intimately tied to the role the phosphate radical plays in the molecule.

The Fitness of Phosphates

THE STORY of the so-called arsenic-eating bacterium highlighted the unique fitness of phosphates for biology and especially for energy transfers. But as Steven Benner pointed out, their unique fitness had been well established for a century before Wolfe-Simon isolated GFAJ-1 from Lake Mono. Their unique fitness for energy transfers was reviewed in 1987 by biochemist F. H. Westheimer in the journal *Science*. "The principal reservoirs of biochemical energy are phosphates," he wrote. "Many intermediary metabolites are phosphate esters, and phosphates or pyrophosphates are essential intermediates in biochemical syntheses and degradations."[15]

In fact, the chemical characteristics of the phosphates are exactly those required for an energy-rich molecule to fund biochemistry in the aqueous medium of the cell. In contrast to other energy-rich compounds, phosphates "can persist in an aqueous environment even though they are thermodynamically unstable, and thus can drive chemical processes to completion in the presence of a suitable catalyst (enzyme)," Westheimer writes. "This remarkable combination of thermodynamic instability and kinetic stability was noted many years ago by Lippmann, who correctly ascribed the kinetic stability to the negative charges in ATP. A citric acid anhydride would not survive long in water and could not serve as a convenient source of chemical energy."[16]

As I elaborated in *Nature's Destiny*, continuing to draw on the work of Westheimer:

> Organic chemists invariably use quite different compounds when carrying out analogous reactions to those carried out by phosphates in living things; for synthetic reactions, chlorides, bromides, iodides, tosylates, triflates, trialkylamines, sulfoxides and selenoxides among others; for activating molecules for reaction, compounds such as carbodiimides.... While these compounds may well suit the organic chemist, living things could not tolerate [such] reactive compounds... as they would inactivate the delicate machinery of the cell.

The energy-rich activating compounds used in the cell are phosphates, such as ATP and GTP. These are far more stable and far less reactive than their equivalents used in organic chemistry, but still sufficiently labile and reactive to fulfill these roles in the cell.[17]

Arthur Needham also stressed the uniqueness of the phosphates for energy transfers. "The reason why these pyrophosphates, and ATP in particular, are the universal mediators of ~P energy is that they have outstandingly high kinetic stability," he writes. ATP, he notes, "hydrolyses very slowly without enzymic help, and it has the best on both counts, energy to give and the power to control flow. The nucleoside moiety is invaluable here since inorganic pyrophosphates [unlinked to adenosine] in fact do hydrolyse in about three minutes at 100°C; this is one more example of the restraining influence of nucleic acids and their components."[18] Peter Atkins concurred, calling phosphorus in ATP a "perfect vector" for carrying out the subtle energy deployments required by biochemistry.[19]

There is another reason why phosphates are strikingly fit for life. The energy levels of the phosphate bond are just about right for its role of imparting energy to enable various chemical processes in the cell. Rob Phillips and his co-authors make the point, noting that ATP works well as an energy currency because the amount of energy released on hydrolysis of the terminal phosphate "is comparable to the energy consumed in many kinds of biochemical transformations and is intermediate between thermal energy... and the energy of a typical covalent bond.... ATP can be considered as the 20 dollar bill of the cell because of its intermediate value in the overall energy economy of the cell. Spending money in large chunks such as 100 dollar bills is unwieldy because they are hard to break. On the other hand, paying with just dollar bills is a nuisance because it takes many of them to buy anything useful."[20]

In four billion years of evolution, the many forms of life on Earth have repeatedly testified to what chemists have recently affirmed: phosphate radicals are the best means of storing and transferring energy for the activities of the cell. In no organisms, not even the most extreme of the extremophiles, has any replacement been found for ATP. This strongly suggests that ATP and energy-rich phosphates are uniquely suited for the roles they fill. Without them, no carbon-based cell, even the most primitive that we can conceive of, could ever have emerged, either on Earth or on some exoplanet in "a galaxy far, far away."

Indeed, as Needham confesses, citing an earlier researcher, Edward O'Farrell Walsh, "It is no exaggeration to say 'without phosphorus: no life.'"[21]

Glycolysis and Respiration

CELLS MAKE ATP in two ways. Virtually all organisms possess an anaerobic metabolic pathway—the glycolytic pathway—which provides energy to generate ATP in the absence of free molecular oxygen.[22] The pathway involves several reactions that convert one molecule of sugar into two pyruvic acid and two ATP molecules. And because pyruvic acid is also readily converted to ethanol (alcohol), the overall chemical process is often referred to as *fermentation.*

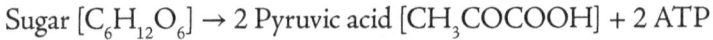

Sugar $[C_6H_{12}O_6] \rightarrow$ 2 Pyruvic acid $[CH_3COCOOH]$ + 2 ATP

As an aside even though the reactions involved in glycolysis do not involve free oxygen, many are classed by chemists as "oxidations." Oxidation refers to an atom or molecule experiencing a net loss of electrons, regardless of whether oxygen is involved. Because of the near universality of glycolysis, in one sense all life on Earth uses a form of oxidation.

Virtually all organisms, including humans, also generate ATP by what is referred to as cellular respiration. This term generally refers to the aerobic process whereby partially oxidized products of glycolysis such as pyruvate are fully oxidized by free molecular oxygen to CO_2 and H_2O, releasing energy used to synthesize ATP molecules.

Pyruvate + $O_2 \rightarrow CO_2 + H_2O$ + 28 ATP

The complete oxidation of sugar to pyruvate and finally to CO_2 and H_2O (glycolysis plus cellular respiration) yields approximately fifteen times the number of ATP molecules produced by glycolysis alone.[23]

In common usage, cellular respiration refers to the aerobic type, which occurs in the mitochondria of all advanced aerobic organisms; but this is not the only type of cellular respiration. Many unicellular organisms use other electron acceptors as the terminal oxidant besides oxygen. Methanogens use CO_2, which is reduced to methane (CH_4). Other microbes reduce sulfate SO_4 to sulfide. Some even "breathe metals," reducing ferric ions Fe^{3+} to ferrous ions Fe^{2+}, while others reduce Mn^{4+} to Mn^{2+}. But all these alternative "oxidants" generate far less energy than the reduction of oxygen to water. All anaerobes using alternative oxidants instead of free oxygen are simple, unicellular lifeforms.

Glycolysis and aerobic cellular respiration are carried out in different compartments of the cell. The reactions of glycolysis occur in the cytoplasm of our body cells while the reactions involved in cellular respiration occur in the mitochondria.

Only cellular respiration, involving the complete oxidation of the body's biofuels to CO_2 and H_2O, provides sufficient ATP to supply the energy demands of advanced metabolically active organisms such as ourselves. It provides energy above and beyond that needed for basic processes such as cell division and the synthesis of basic material constituents of the cell. This makes it possible to do interesting things and grow in complexity beyond the basic need to survive.

More than seventy years ago, George Wald noted the vital importance of the additional energy that respiration provides for complex life using oxygen as a terminal electron acceptor. "It is difficult to overestimate the degree to which the invention of cellular respiration released the forces of living organisms," he wrote in a well-known *Scientific American* article. "No organism that relies wholly on fermentation [glycolysis] has ever amounted to much.... Respiration used the material of organisms with such enormously greater efficiency as for the first time to leave something over.... To use an economic analogy, photosynthesis brought organisms to subsistence level; respiration provided them with capital. It is mainly this capital that they invested in the great enterprise of organic evolution."[24]

Thus, bees buzz and hummingbirds hum, squids and chameleons change color, amoebas engulf prey, crows solve problems, and humans build rockets to the stars, only because oxidation releases metabolic energy in quantities much greater than are needed for merely sustaining the basic metabolism of the cell.

Proton Pumping

THE MECHANISM by which cells use cellular respiration to manufacture ATP is not only one of the wonders of cell biology, but also one of the most unexpected and important discoveries of twentieth-century science. The mechanism, described by the chemiosmosis hypothesis, was first proposed by British biochemist and Nobel laureate Peter Mitchell in a letter to *Nature* in 1961,[25] but was not thoroughly validated and widely accepted until the following decade.[26]

The mechanism, which origin-of-life researcher Leslie Orgel called "the most counterintuitive idea in biology since Darwin,"[27] involves exploiting the energy discharged by electrons flowing down an energy gradient to pump protons through a bilayer lipid membrane. These, according to Mitchell's model, then flow back through the same membrane, providing energy for the synthesis of ATP,[28] which is carried out by an extraordinary rotating enzyme, ATP synthase, one of the most complex of all known enzymes.

As Nick Lane puts it:

Essentially all living cells power themselves through the flow of protons (positively charged hydrogen atoms), in what amounts to a kind of electricity—proticity—with protons in place of electrons. The energy we gain from burning food in respiration is used to pump protons across a membrane, forming a reservoir on one side of the membrane. The flow of protons back from this reservoir can be used to power work in the same way as a turbine in a hydroelectric dam. The use of cross-membrane proton gradients to power cells was utterly unanticipated. First proposed in 1961 and developed over the ensuing three decades by one of the most original scientists of the twentieth century, Peter Mitchell, this conception has been called the most counterintuitive idea in biology since Darwin, and the only one that compares with the ideas of

Einstein, Heisenberg and Schrödinger in physics... the use of proton gradients is universal across life on earth—proton power is as much an integral part of all life as the universal genetic code.[29]

Before Mitchell proposed it, no one imagined that ATP synthesis would involve such a mechanism.

The pumping of the protons across the membrane is empowered by electrons released from the metabolic breakdown of basic foodstuffs like sugars and lipids (or from inorganic minerals in the case of chemosynthetic bacteria) flowing down an energy gradient along what are known as electron transport chains (ETCs) to a terminal electron acceptor.

The energy gradient along the ETC is caused by the strong pull of the terminal electron acceptor for electrons (oxygen in animal and plant mitochondria), analogous to the pull of gravity when water flows downhill. And in the case of aerobes using oxygen as the terminal electron acceptor, oxygen's "desire" for electrons is so great that the gradient is very steep. And just as the tumble of water down a mountainside can be used to grind corn in a water mill or generate hydroelectric energy, the energy gradient of the electron transport chain can be exploited to do metabolic work.

Efficient use of the energy discharged by the flow of electrons along the ETC depends, however, on the electrons flowing down the gradient in a series of small discrete steps, analogous to capturing hydro energy from the flow of a river in a succession of small water mills. If the energy were instead discharged in one large, uncontrolled jump, much of it would be wasted and of little use to the cell, like hydro energy discharged when water falls forcibly and uncontrollably over a waterfall.

As the authors of *Molecular Biology of the Cell* explain:

Because of the huge free-energy drop, this reaction [$2H^+ + 2e^- + \frac{1}{2}O_2 \rightarrow H_2O$] would proceed with almost explosive force and nearly all of the energy would be released as heat. Cells do perform this reaction, but they make it proceed much more gradually by passing the high-energy electrons from NADH to O_2 via the many electron carriers in the electron-transport chain. Since each successive carrier in the chain holds its electrons more tightly, the highly energetically favorable reaction $2H^+ + 2e^- + \frac{1}{2}O_2 \rightarrow H_2O$ is made to occur in many small steps. This enables nearly half of the released energy to be stored [and available for work], instead of being lost to the environment as heat.[30]

Electron Transport Chains

ELECTRON TRANSPORT chains (ETCs), which carry the electrons down the energy gradient, are present in virtually every life form on Earth and in all life forms synthesizing ATP by proton flows across membranes. And it is widely assumed that electron flows in ETCs coupled to the generation of proton flows—or "proticity" as Lane puts it—provided the energy for ATP synthesis in the very first life forms on Earth.[31]

The ETCs in all nucleated cells are embedded in the inner mitochondrial membrane, while in bacterial cells they are embedded in the plasma membrane, which bounds the bacterial cell. Like all bio-membranes, the mitochondrial and bacterial cell membrane share the same lipid bilayer structure, itself a marvel of bioengineering fitness, as we saw in a previous chapter.

ETCs are made up of several protein complexes, within which are a series of what are termed iron-sulfur clusters. Most of these consist of four iron and four sulfur atoms linked together in what is essentially an inorganic crystal.[32] There are nine sulfur clusters in the mitochondrial ETC, arranged about fourteen angstroms (Å) apart in a sequence leading through the protein complexes.[33] The electrons released from the metabolism of food stuffs enter the chain and hop from one cluster to the next by quantum tunneling. They finally end up at the terminus—the enzyme cytochrome c oxidase in aerobic cells—where they reduce oxygen to water.[34]

Proticity

As EACH individual electron hops from one iron-sulfur cluster to the next along the ETC, it releases energy, which is used to pump an individual proton across the inner mitochondrial membrane (or the plasma membrane in bacteria), building up the concentration of protons on the other side. And it is this buildup that creates the proticity that, flowing

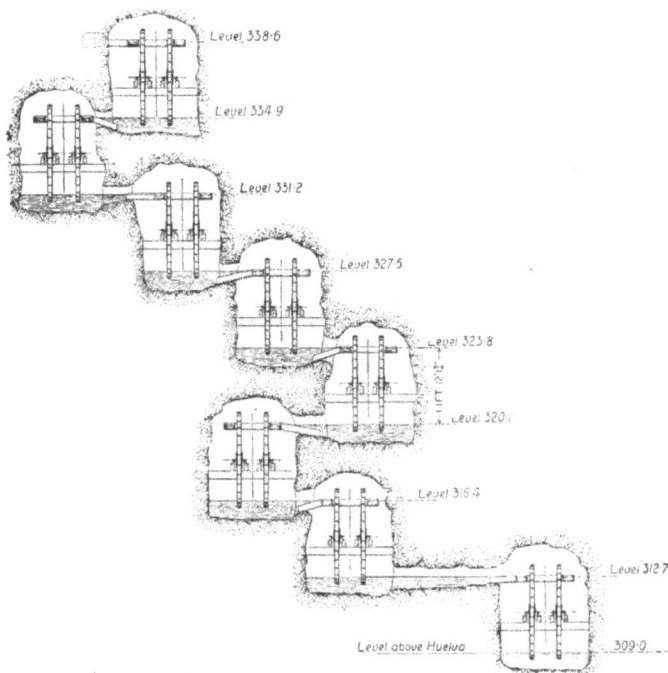

Figure 5.3. Ancient Roman drainage system at Rio Tinto mines, Andalusia, Spain. A graphic example of energy expended in a number of small steps.

back across the membrane, empowers ATP synthesis—about three protons for each ATP molecule synthesized.[35]

A wonderfully vivid picture of individual electrons hopping "from centre to centre" down the ETC is given in Lane's *The Vital Question*: "The regular spacing of these centres [the iron-sulfur clusters] suggests that they 'tunnel' by some form of quantum magic, appearing and disappearing fleetingly, according to the rules of quantum probability," he writes, adding that "so long as each redox centre is spaced within about 14 Å of the next, and each one has a slightly stronger affinity for an electron than the last, electrons will hop on down this pathway of redox centres [the iron-sulfur clusters] as if crossing a river on nice regularly spaced stepping stones... drawn onwards by the powerful tug of oxygen [and] its voracious chemical appetite for electrons." Or to use another word picture, it is "a wire, insulated by proteins and lipids, channeling the current of electrons from 'food' to oxygen."[36]

Lane continues describing additional details about this marvel of biomolecular fitness. "The electrical current animates everything here," he writes, and the giant protein complexes are "full of trip switches."[37] Lane's subsequent description is worth quoting at length:

> If an electron sits in a redox centre, the adjoining protein has a particular structure. When that electron moves on, the structure shifts a fraction, a negative charge readjusts itself, a positive charge follows suit, whole networks of weak bonds recalibrate themselves, and the great edifice swings into a new conformation in a tiny fraction of a second. Small changes in one place open up cavernous channels elsewhere in the protein. Then another electron arrives, and the entire machine swings back to its former state. The process is repeated tens of times a second.... [P]rotons bind to immobilised water molecules, themselves pinioned in their place by charges on the protein.... these water molecules shift when the channels reconfigure themselves... [and the protons] are passed from one water molecule to another through dynamic clefts, opening and closing in swift succession, a perilous route through the protein that slams closed instantly after the passage of the proton, preventing its retreat.... All that power, all that ingenuity, all the vast protein structures, all of that is dedicated to pumping protons across the inner mitochondrial membrane. One mitochondrion contains tens of thousands of copies of each respiratory complex. A single cell contains hundreds or thousands of mitochondria. Your 40 trillion cells contain at least a quadrillion mitochondria, with a combined convoluted surface area of about 14,000 square *metres*; about four football fields. Their job is to pump protons, and together they pump more than 10^{21} of them—nearly as many as there are stars in the known universe—*every second.*[38]

Transition Metals

The conduction of electrons along the ETCs in a stepwise series of jumps from carrier to carrier in the electron transport chain depends on one of the unique properties of the transition metals atoms: their capacity to accommodate different numbers of electrons in their outermost electron shells, to possess what are known more formally as different oxidation states (or redox potentials).

One consequence of this is that their affinity for electrons (their oxidation states) will vary depending on how many electrons are accommodated in their outer electron shells. When the number is high the atom will have a low affinity for electrons; when the number is low they will have a high affinity for electrons.

Moreover, the affinity for electrons of a transition metal (its oxidation state) can be, as Robert Crichton points out, fine-tuned by subtle changes in its immediate biochemical environment in the cell.[39]

And it is primarily because their affinity for electrons can be fine-tuned to several different oxidation states that it is possible to arrange a chain of transition metal atoms of increasing incremental affinity for electrons to draw them along the ETC in a series of discrete steps.

Robert J. P. Williams, one of the world's leading twentieth-century authorities on the role of metals in biology, commented in a classic paper on the fitness of transitional metal atoms for the design and construction of electron transport chains:

> In this article I shall take it as demonstrated that electron flow is over-whelmingly from metal centre to metal centre, that these centres are often entatic, i.e. they are structurally designed for this purpose, that they are placed at about 15 Å apart to allow adequate rates of electron transfer and that their positioning is such as to make directional local circuits... Protein side-chains play a minor role in the electronic circuits of biology. They just hold and adjust the metal ion energies while permitting the minimum required mobility.[40]

And as Williams continues, "Easy electron transfer is a general property of transition metals but not of proteins." He speculates that primeval life forms may have used transition metals for energy capture "before there was protein synthesis." He concludes "that transition metal inorganic elements in some organisation are of the essence of life as much as this is true of amino acids and nucleotides."[41]

Only the transition metal atoms possess precisely the properties required to form an electronic circuit in which the electrons will lose their energy in discrete ordered steps. No other atoms will do. And intriguingly, it is because of their unique electrical conducting properties that the transition metals are used in human technology to make wire conductors. As Williams and J. J. R. Fraústo da Silva put it, "Man makes his wires from metals such as copper; biology makes hop conductors from metal ions embedded in... protein."[42]

Although many of the transition metals possess similar properties, iron is the most important for life. Life without iron is almost as inconceivable as life without carbon. "When we invoke the extensive range of redox potentials available to the metal by varying its interaction with coordinating ligands and add to that its capacity to participate in one electron transfer (i.e. free radical) reactions," Crichton writes, "it is easy to see why iron is virtually indispensable for life."[43]

Four Billion Years

As FAR as current life on Earth is concerned, all advanced organisms and the vast majority of microbes manufacture ATP by exploiting the discrete quanta of energy released as electrons race towards their destiny at the terminus of the chain. In all organisms the electron flow in electron transport chains is along conducting wires made up of transition metal atoms. And virtually all organisms use the energy released as the electrons jump from metal atom to metal atom to pump protons across a bilayer lipid membrane. Moreover, to my knowledge, no biochemist has ever proposed a theoretical alternative to the use of the transition metals to form the conducting wires.

In four billion years, no cell, not even the most exotic extremophile, has discovered any alternative mechanism for extracting the energy of metabolism in discrete quanta to generate the vast quantities of ATP necessary to support life. Nor has any cell found any replacement for the transition metal components of the ETC. Similarly, in all organisms, energy-rich phosphates are used for transferring energy and promoting chemical reactions. Again, it has been so since the beginning. There is no other choice.

If the transition metals did not have precisely the properties they possess—readiness to accept and donate electrons and to exist in multiple oxidation states—and if ATP did not have precisely the properties it does, they would have to be invented. For without their unique fitness for controlled energy generation and exploitation, carbon-based life would be a far more mundane phenomenon, restricted to simple unicells depending on glycolysis for energy production.

In short, this chapter has provided further evidence of the striking fitness of nature for life as it exists on Earth, conveying the powerful impression that a blueprint for the production and utilization of energy by carbon-based cells was written into the laws of nature from the moment of creation.

6. No Biology
Without Metals

The effort to grasp the world as a single unified whole
runs through all the medieval summae, the encyclopedias
and the etymologies... The philosophers of the twelfth
century speak of the necessity of studying nature: for
in the cognition of nature in all her depths, man finds
himself... underlying these arguments and images is
a confident belief in the unity and beauty of the world,
and also the conviction that the central place in the world
which God has created belongs to man.
—Aron Gurevich, Categories of Medieval Culture[1]

IRON'S SIGNIFICANCE TO BIOLOGY BEGAN LONG BEFORE LIFE FIRST
emerged in the primeval oceans. Long before mankind developed met-
allurgy and exploited the properties of this most useful of metals, and
long before its unique properties were exploited to carry oxygen in the
blood, they played a vital role in crafting Earth into a life-friendly habi-
tat. Iron atoms, drawn by gravity to the center of the primeval Earth,
are thought to have generated much of the heat that caused the initial
chemical differentiation of the Earth into crust, mantle, and core. Without this differentiation, the vital recycling of the crustal materials in the
tectonic system would never have begun. The same heat contributed to
the outgassing of the early atmosphere and ultimately the formation of
the hydrosphere.

And long before the Earth existed, even before there was a solar system, iron was necessary. The accumulation of iron atoms in the center of

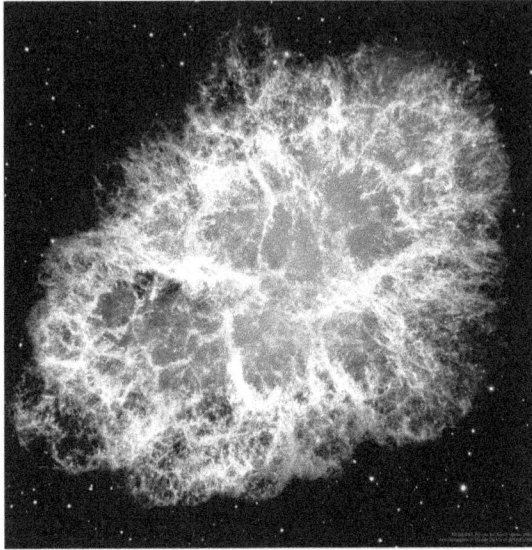

Figure 6.1. The Crab Nebula, the remains of a supernova that exploded in 1054.

high-mass stars triggers the explosion of supernovae, such as that in the Crab Nebula (see Figure 6.1), which distribute the atoms of the periodic table throughout the cosmos and make them available for the formation of solar systems, the Earth, and ultimately life.

Iron (Fe) is the most abundant element in the Earth, making up about 30% of the Earth's mass. The core of the Earth is a vast ball of molten iron, and iron is the fourth most common element in the Earth's crust.[2]

Every day since life first emerged in the primeval ocean, iron has been acting as an unseen guardian. Molten iron in the Earth's core is thought to act like a gigantic dynamo, generating the Earth's magnetic field. This in turn creates the van Allen radiation belts that shield the Earth's surface and all life on the Earth's surface from destructive, high-energy, penetrating cosmic radiation as well as preserve the crucial ozone layer from cosmic ray destruction. Without the iron atom, there would have been no heating of the primitive Earth, no tectonic recycling and uplift, no atmosphere, no hydrosphere, no van Allen radiation belts, no protective magnetic field, no hemoglobin, no oxidative metabolism, no electron transport chains (ETCs), no advanced life forms, and possibly no life at all.

Our understanding of the important role of iron and other metals in biochemistry and cell biology has increased enormously since Lawrence Henderson wrote his classic *Fitness of the Environment* in 1913. At that time, virtually nothing was known about the role of metals in living systems. And even as recently as a few decades back, knowledge in this field was so limited that Nobel laureate Sir Hans Krebs was able to comment that, for all he knew, most metal ions found in biology would turn out to be damaging impurities.[3]

We now know that Krebs was dead wrong. Metals play a vital role in living systems—so vital that one of the leading twentieth-century experts in this field, Professor Robert Williams, in a landmark review entitled "The Symbiosis of Metals and Protein Function," concluded that "biology without metal ions does not exist any more than biology exists without DNA or proteins." As he went on to say, "The machinery of life rests with these two components, metal ions and proteins... The all-pervading influence of metal ions in biological systems is such that I now declare that in my mind there is no biology without metal ions."[4]

It is hard to disagree with Williams when close to one-third of all enzymes involve a metal ion as an essential participant[5] and when, as we saw in Chapter 5, the transitional metals iron and copper play indispensable roles in electron transport chains. Also, the alkali metal ions Na^+ and K^+ play critical roles in maintaining the membrane potential. Todor Dudev and Carmay Lim echoed Williams in an article in *Chemical Reviews*:

> Metal ions are required for growth of *all* life forms. Currently, about half of all proteins contain metal ions, and most ribozymes [RNA molecules with enzymatic functions] *cannot* function without metal ion(s). Metal ions perform a wide variety of specific functions associated with life processes. One function uniquely performed by metalloproteins is respiration, whereby an iron center in the hemoglobin-myoglobin fam-

ily and hemerythrins or a copper center in hemocyanins binds an oxygen molecule reversibly. In many cases, metal ions, e.g., Zn(II), Mg(II), Ca(II), stabilize the structure of folded proteins, while in other cases they help to fix a particular physiologically active conformation of the protein. Metal ions are an integral part of many enzymes and are indispensable in several catalytic reactions... In particular, transition metals, such as Fe, Cu, and Mn, are involved in many redox processes requiring electron transfer. Alkali and alkaline earth ions, especially Na(I), K(I), and Ca(II), play a vital role in triggering cellular responses.[6]

One of the fatal "defects" of a "biology without metals" is, as Williams points out, that the properties of different amino acids do not vary greatly when incorporated into proteins, while the metal ion brings to proteins a great diversity of chemical properties and molecular geometries that greatly enrich their physio-chemical properties and consequent catalytic abilities.[7]

Robert Crichton sounds a similar note in a reflective introductory overview of his lucid *Biological Inorganic Chemistry*. As evidence of the crucial importance of the metals, Crichton cites the need to transport electrons in electron transport chains (ETCs).[8] Transitional redox metals such as iron and copper are, as he points out, far better at doing this than organic compounds such as flavins. We have seen in the previous chapter that electron flows down ETCs are universally conducted by transitional metal atoms. Without ETCs and the metal conductors they use, the ability of cells to generate bio-energy would be greatly constrained and only the simplest types of unicellular life would be possible. And as also mentioned in the previous chapter, in four billion years, no organism has managed to replace the electron-conducting functions of iron and copper with any organic nonmetal compounds.

Crichton offers several additional examples of the essential role of metals in biology. "The water-splitting centre of green plants (photosystem II), which produces oxygen, is based on the sophisticated biological use of manganese chemistry,"[9] he writes. He also notes that "the enormous negative charges that are generated along the polyphosphate backbone of nucleic acids need to be balanced by positively charged counterions."[10] Many of the vital functions of cells depend on the amplification of "signals, arriving at the cell membrane in nanomolar concentrations, but which result in millimolar intracellular responses." For the transmission of electrical signals, as in the nervous system of higher organisms, it's essential that cells are able to "generate transmembrane electrical potentials."[11] And perhaps most important of all, he adds, metals are needed as cofactors to "enable the proteins which we call enzymes to catalyse reactions, many of which would quite simply be impossible if we relied solely on organic molecules."[12]

Crichton concludes, "For almost all of these purposes, large, cumbersome and bulky proteins are clearly not the answer."[13] All these tasks require metallic elements in addition to the six foundational elements carbon, hydrogen, oxygen, nitrogen, phosphorus, and sulfur.

As a fascinating aside, it is intriguing to note that Henderson, fifty years before Krebs, saw the metals in biological systems as having utility, rather than as being impurities. He wrote, "The physiological utility of compounds containing the elements of inorganic chemistry is very great,"[14] and he noted that "haemoglobin contains iron, and the capacity of haemoglobin to unite with oxygen… is unquestionably due to the chemical behavior of that metal."[15] This proved correct, as did his insistence that "copper in the… haemocyanines, fulfill[s] a similar function in lower animals."[16]

So Henderson the teleologist assumed that the metal atoms were of utility fifty years before Krebs. Henderson was also fifty years ahead of Fred Hoyle, who is widely credited with having made the first "anthropic prediction" by proposing the existence of special resonances of the carbon nucleus, which allowed the atom-building process to proceed to the higher elements in the stars.

The spectacular fitness of the metals for specific biological functions was reviewed in Williams's 1985 paper (cited above) and has been systematically reviewed in several recent books.[17] We now know that at least ten different metal atoms—sodium, potassium, magnesium, calcium, cobalt, copper, iron, manganese, molybdenum, and zinc—play essential roles in the cell,[18] are essential in the human diet, and are uniquely fit to serve very specific biochemical functions. Without their exact properties, carbon-based life on Earth could never have arisen.

Individual metal atoms are fit for very different biological functions because each metal atom has properties that confer unique capabilities, which enable each to play specific, vital roles in biochemistry. And like the different properties of carbon, hydrogen, oxygen, and nitrogen, the differing chemical and physical properties among the metal atoms are critical ingredients of fitness in nature for life. While the metals do not vary one from another to the same degree, they do exhibit critical differences that enable them to fulfill very different roles in the cell.

The pages that immediately follow review key properties of some of the essential metal atoms, properties that make them uniquely fit for various biochemical functions. To keep the text more accessible to non-specialists, I will place much of the technical material for this chapter in endnotes.

Sodium and Potassium

BECAUSE SODIUM (Na) and potassium (K) bind only weakly to organic compounds, they are highly mobile and ideally suited to move electric charge at great speed, create electronic gradients across cell membranes, and ensure the maintenance of charge balance on both sides of the cell membrane.[19]

As mentioned in Chapter 4, their speed is indeed remarkable. "Up to 100 million K^+ ions per second may pass through one ion channel in the cell membrane—a rate 10^5 times greater than the fastest rate of transport mediated by any known carrier protein," notes one widely used cell biology textbook.[20] Potassium and sodium are the only two ions whose speed is unimpeded by a tendency to bind to organic compounds. No other ions are available for this vital function. Their ability to carry charge across the cell membrane at such speed is essential for nerve impulse transmission along the axons of the nerve in higher organisms. Without it there would be no central nervous system, which is essential to humans and other advanced organisms.

In four billion years of evolution, life has found no alternative compounds or atoms to replace the bio-functions of these two ions. It is no exaggeration to say that our intelligence and conscious self-awareness is gifted in part by these two tiny, highly mobile, charge-conveying metal ions.

Calcium

THE PHYSIOLOGICAL functioning of the cell requires rapid information transmission by a chemical messenger to trigger particular protein functions by causing sudden conformational changes. One of the best examples of this is the release of calcium ions from special vesicles in muscle cells, which triggers muscle contraction.[21]

Calcium (Ca) along with sodium (Na), potassium (K), and magnesium (Mg) provide the cell with a set of highly mobile ions. Although all of these ions (Na^+, K^+, Ca^{2+} and Mg^{2+}) are highly mobile, a chemical messenger needs to have properties besides just rapid movement. It must also be able to bind to another molecule in the cell—generally a protein—with high specificity to enact a specific conformational change. The necessity for high-affinity binding largely excludes the monovalent metal ions sodium and potassium, leaving calcium and magnesium to fulfill this essential messenger work. And of these two, calcium is superior.

As Williams comments, "Nearly all triggering of muscle and similar mechanical actions is due to calcium. Now amongst the metal ions available to biology only calcium can be high in concentration, can diffuse rapidly, and can bind and dissociate strongly and yet binds strongly."[22] Of the two divalent ions, calcium and magnesium, the binding of calcium to most protein sites is many times stronger than magnesium,[23] and hence calcium is far fitter for this role.

Consequently, in biological systems, calcium is primarily used where chemical information must be rapidly transmitted to a target protein to trigger a particular cellular function. This can be muscle contraction, transmission of nerve impulses across the synapse, hormone release, or changes following fertilization.[24]

A fascinating reciprocal element of fitness in the calcium-protein relationship is the capacity of the alpha helix, one of the basic protein structural subunits, to react rapidly to the stimulus of calcium binding. Williams, commenting on the suitability of the helical structures in proteins to respond rapidly to the stimulus of calcium, remarks:

> To make a sensible system the proteins which are in the muscle or the internal filamentous units of cells must have activity matching that of calcium and this means it is their fast dynamics as much as their structure which have evolved. These proteins are largely based on helices. In a general sense a helical rod is useful in that its movement economically connects activities at either end through rotational/translation movement like that of a screw or a worm-gear. It is fast since helix/helix movement need not break H bonds. We see in a helix the potential for matching the dynamics of the calcium ion.[25]

The unique fitness of calcium for its role as chemical messenger in the cell is highlighted by its universal use for this purpose. It is the same in all cells. No other organic compound or alternative metal atom has been substituted for the role of calcium in any living cell on earth in billions of years of evolution.

Magnesium

Magnesium (Mg) is present in every cell type in every organism. Magnesium enzymes are necessary for the catalytic action of more than 300 enzymes.[26] Many of the enzymes involved in intermediary metabolism depend on magnesium, as do many enzymes involved in nucleic acid metabolism.[27]

Although magnesium is calcium's homologue in period 3 of the periodic table, its chemical and physical properties differ in various subtle ways that equip it for very different roles in the cell.[28] As mentioned above, the magnesium ion is unsuitable to take the place of calcium as a cellular messenger to trigger conformational changes in a target protein, because the Mg^{2+} ion tends to bind to proteins far less strongly than calcium.[29] But for certain functions, magnesium would seem to be uniquely fit.

ATP: It is hard to think of a biological function more universal and important than that carried out by ATP, the main provider of cellular energy. The cell's ability to use the energy released on the hydrolysis of ATP and other energy-rich phosphates such as guanosine triphosphate (GTP) depends on the association of the nucleotide triphosphate with a magnesium ion.[30] What is called ATP is actually "Mg^{2+}-ATP," and this appears to have been the case for billions of years from the time ATP was adopted as the energy currency of all life on Earth. In all that time, no other metal ion has replaced the role of Mg^{++} in any known cell. A detailed mechanistic explanation is given in many texts,[31] but on the empirical evidence alone it is reasonable to think that the magnesium atom is uniquely fit for its role.

Chlorophyll: For oxygen-utilizing aerobic life forms, magnesium plays a vital role in the main light-collecting molecule, chlorophyll. Peter Atkins waxes lyrical about its role in the capture of sunlight:

> Without chlorophyll, the world would be a damp warm rock instead of the softly green haven of life that we know, for chlorophyll holds its magnesium eye to the sun and captures the energy of sunlight, in the first step of photosynthesis…. magnesium has exactly the right features to make this process possible. Had the kingdom lacked this element, chlorophyll's eye would have been blind, photosynthesis would not take place, and life as we know it would not exist.[32]

Atkins speaks the truth. Although photosynthesis is not essential to life (since living systems can derive energy from sources other than the radiant energy of a star) stellar radiation is nevertheless the only reliable, evenly distributed source of energy available on planetary surfaces. It is difficult to conceive of a biosphere as rich as our own, with diverse and complex life forms, on any planetary surface that does not draw on the radiant energy of its sun. And for that, magnesium holds the key.

The magnesium ion can be replaced by manganese, iron, cobalt, nickel, copper, and zinc in various metal porphyrins,[33] but none can mimic the light-absorbing capacity of magnesium. This is due to certain unique features of the magnesium ion.[34] As Melvin Calvin remarked, there must be "something very special about the electronic structure of the magnesium" ion that confers such a remarkable fitness for light capture on chlorophyll. As he points out, the light-absorbing power of magnesium chlorin is "several thousand times that of iron porphyrin."[35]

Williams elaborates:

> The effect of incorporating any metal ion into a chlorin is to change its rough twofold symmetry into a rough fourfold symmetry. The chlorin visible absorption spectrum then changes from a series of four well-spaced absorption bands… to effectively a single intense band at the longest wavelength and thereby magnesium chlorophyll provides an excellent lowest energy, light-capture, device. It is the choice of both chlorin and magnesium which generates the property. Now any metal ion of the group Mg^{2+}, Zn^{2+}, Ni^{2+}, Cu^{2+}, Mn^{2+} and Co^{2+}, would cause roughly the same spectrum to appear but Mg^{2+} is chosen despite its weaker binding because it is lighter than any of the others and consequently generates very little fluorescence, i.e. loss of energy.[36]

Magnesium captures light energy with remarkable efficiency, approaching the theoretical limit for conversion of light into electrical energy. In a *Nature* article on the topic, the authors commented on the efficiency of the photosynthetic apparatus: "The yield of the process is close to its maximum value, unity—an achievement that remains unmatched in model systems."[37]

The Fitness of the Transition Metals

The transition metal atoms—including manganese, iron, cobalt, copper, zinc, and molybdenum—have their own set of indispensable properties upon which the life of the cell depends. As mentioned earlier in the book, their outer shell and next innermost shell can contain variable numbers of electrons. And this means they can possess different oxidation states and different affinities for electrons. As we saw, this enables iron and copper in particular to function as electron carriers in electron transport chains.

Moreover, because they have different oxidation states, the transition metals are uniquely fit to channel electrons down gradients in discrete energy steps, allowing for the efficient harnessing of the energy released to perform chemical work. Yet again, in four billion years, no organism has found any organic compounds or alternative metal atoms to substitute for the transition metals in ETCs. And this implies that if nature had not provided these metals, with their particular suite of properties, then the controlled and efficient extraction of energy from metabolic processes would never have been achieved.

The transition metals contain what are called "unpaired electrons," and this endows them with another unique ability—to donate one electron at a time to activate oxygen for chemical reaction. Donating one electron at a time overcomes what is termed the "spin restriction" of the oxygen atom (a unique property which greatly attenuates the chemical reactivity of oxygen at ambient temperatures) and prepares the oxygen atom for chemical reaction. And for this reason it is always transition metals, most importantly iron and copper, which are used in enzymes involved in oxygen activation, transport, and storage.[38] Without their unique capacity to activate and reduce oxygen, oxidative metabolism would be impossible. To paraphrase Williams, there is no oxidative metabolism without the transition metals.

Hemoglobin

AEROBIC METABOLISM in large life forms like ourselves depends on the transport of oxygen in the bloodstream from the lungs to the tissues where the energy of oxidation is released and used to generate the chemical energy carrier ATP. The carrier molecule in all complex organisms is always a transition metal in a loose reversible combination with oxygen.[39] In the blue blood of an octopus, oxygen is reversibly bound to copper in hemocyanin, while in the red blood of mammals, oxygen is bound reversibly to iron in hemoglobin.

Figure 6.2. Rust on chain links near the Golden Gate Bridge in San Francisco.

This basic empirical observation hints strongly that the transition metals must possess some special fitness for reversibly bonding to oxygen. Why else, in billions of years of evolution, has no organism come up with a way to transport oxygen without transition metals?

As is well known, hemoglobin is the carrier molecule not only in mammals but in all other vertebrates. It is one of the most familiar of all biological molecules and the substance that makes blood red.

What is curious about iron's ability to bind gently and reversibly with oxygen in hemoglobin is that, despite the kinetic barriers to oxygen activation imposed by the spin restriction, oxygen does combine slowly with other atoms even at ambient temperatures, binding with them very strongly and irreversibly. Butter turns rancid, iron rusts, and carbon is oxidized to carbon dioxide.

But rust (Fe_2O_3) does not easily release its oxygen, which can be extracted only by heating the iron oxide to 1,500°C. Clearly, the chemical state of iron in the cables of the Golden Gate Bridge—before its union with oxygen—is different in some significant ways from the chemical state of iron in hemoglobin. One major difference is that the iron in the hemoglobin is not "free iron" but is part of a porphyrin-iron group, "heme," that sits in a hydrophobic pocket in the hemoglobin molecule where the iron is already bonded via five "coordinate bonds" with nitrogen atoms (four in the porphyrin ring and another in one of the amino acid side chains of the hemoglobin). This leaves one free for binding to oxygen.

The attenuating influence of the coordinate bonds and the special micro-environment of the hydrophobic pocket is indicated by the fact that iron bound in a porphyrin complex outside the protein hemoglobin *does* react spontaneously and irreversibly with oxygen, just like the "free" iron in rusting Golden Gate Bridge cables (Figure 6.2). As the authors of *Bioinorganic Chemistry* comment, "In contrast to *Hb* [hemoglobin] and *Mb* [myoglobin, a molecule somewhat similar to hemoglobin that occurs in muscles], most simple iron(II)-porphyrin complexes that are not protected by a protein environment react irreversibly with O_2 forming oxo-bridged dimers."[40]

The fine-tuning of oxygen binding is further evidenced by the fact that some additional factors, such as high carbon dioxide levels and low pH, cause subtle chemical changes in hemoglobin, which weaken the iron-oxygen bond and promote the release of oxygen in the tissues. Conversely, a high pH or low CO_2 favors stronger binding of the oxygen to the iron atom.[41]

Clearly, the coordinate bonds the iron makes with adjacent atoms in the hemoglobin pocket in conjunction with the special chemical environment in the hemoglobin pocket greatly attenuate the strength of the iron-oxygen bond, and the positioning of certain of the amino acid side chains in hemoglobin are critical. The iron in the bridge cables has no such partners or special chemical micro-environment to attenuate its chemical reactivity.[42] As in so many other instances, Williams notes, "many of the functions of the metal ions could not be induced in purely inorganic materials but had to await the synthesis of their organic molecular partners, such as the porphyrins."[43]

As mentioned above, iron and other transition metals contain unpaired electrons,[44] which can be transferred one at a time to the O_2 molecule, commencing its reduction and activation. After the oxygen ligand is attached to the iron by the remaining sixth coordinate bond, one electron is drawn by the O_2 from the iron, causing its partial reduction to superoxide radical O_2^- and leaving the iron oxidized. This negatively charged highly reactive superoxide radical forms a complex with the positively charged iron.[45] Curiously, it is as this highly reactive and unstable superoxide radical that oxygen is transported in the hemoglobin to the tissues. And the instability of the iron-superoxide complex is one of the reasons for the weakness and ready reversibility of the bonding of iron and oxygen in hemoglobin. The necessary stabilization of the superoxide is assisted, as Crichton points out, "by hydrogen bonding to the distal histidine proton."[46] This is another example of how the micro-chemical environment constrains and, in this case, stabilizes the oxy-hemoglobin.

Further transfers of electrons to O_2^-, which occur in other heme—iron enzymes such as cytochrome P 450 and cytochrome c oxidase, cause further reduction, generating various highly reactive free oxygen radicals. For example, the addition of a second electron gives peroxide (O_2^{2-}), which accepts two protons to give H_2O_2 (hydrogen peroxide).[47] The acceptance of an additional electron by superoxide in hemoglobin is presumed to be inhibited by the stabilizing effect of the distal histidine. "Histidine is valuable as a distal amino acid residue because of its basic nature, which keeps protons away from the coordinated O_2," Kaim and his colleagues write. This guardian role aids stabilization because, as they further note, "protons act as electrophilic competitors in relation to the coordinating iron, weakening its bond to O_2 and thus favoring deleterious autoxidation processes."[48] Additionally, the heme itself may have special properties that promote the reversible binding and stabilization of the superoxide radical.[49]

However, the reversible binding that is achieved is close to a miracle. Recall how tightly oxygen binds to most atoms or molecules outside of hemoglobin. Think of H_2O, CO_2, or iron oxide (Fe_2O_3). Extracting the oxygen from those compounds requires extreme physical or chemical procedures, while oxygen in the hemoglobin comes off effortlessly in the tissues merely because of a drop in the concentration of O_2 molecules.

Without the unique properties of this most important of all transition metals—its coordinate chemistry and ability to commence the reduction of oxygen—our energy-hungry tissues couldn't receive the oxygen they need.

Cytochrome c Oxidase

ONE OF the most important of all enzymes that uses the oxygen-handling capabilities of the transition metals is cytochrome c oxidase. This terminal member of the electron transport chain (ETC) in the mitochondria performs the critical final reaction of oxidative metabolism. This involves transferring the electrons flowing down the ETC to atoms of O_2, reducing them to water. It sits astride the inner bilayer lipid membrane in the mitochondria.

On the significance of cytochrome c oxidase, Earl Frieden comments:

> If a biochemist is asked to identify the *one* enzyme which is most vital to all forms of life, he would probably name cytochrome C oxidase. This is the enzyme, found in all aerobic cells, which introduces oxygen into the oxidative machinery that produces the energy we need for physical activity and biochemical synthesis.... This enzyme may be regarded as the ultimate in the integration of the function of iron and copper in biological systems. Here in a single molecule, we combine the talents of iron and copper ions to bind oxygen, reduce it with electrons from other cytochromes in the hydrogen electron transport chain and, finally, to convert the reduced oxygen to water.[50]

The electrons flow within the molecule along a "transition metal wire" composed of a succession of iron and copper atoms which conducts them to the final catalytic center where they cause the reduction of oxygen to form water.[51] Here we see, in one vital protein complex, the unique ability of transition metals to draw electrons along an energy gradient and adroitly donate electrons one at a time to reduce oxygen via a set of partially reduced intermediates to water.[52] And among the myriad of different organisms that use this key enzyme, none have found any other metals or organic compounds to replace the roles of iron and copper.

Cytochrome c oxidase also contains two other metal atoms—zinc and magnesium. Their roles are not directly involved in the transfer of the electrons in the molecule or in the reduction of oxygen to water. Zinc appears to play a structural role away from the active site, while magnesium may be involved in the release of the water molecules at the active site.[53]

Thus the work of this remarkable nano-machine, whose basic structure is built up of carbon, nitrogen, oxygen, sulfur, and hydrogen, depends on the unique properties of four metal atoms: iron and copper, which are involved in its core function, and zinc and magnesium, in supporting roles. In other words, this atomic machine, central to life, exploits the unique chemical and physical properties of nine of the ninety-two naturally occurring elements.

Finally, it is worth recalling that hemoglobin and cytochrome c oxidase are only two of a vast inventory of enzymes that use iron or another transition metal at their active sites.

Manganese

THE EXTRAORDINARY fitness of the transition metals for using oxidation energy would be to no avail unless oxygen was freely available in great quantities in the atmosphere. And for this, nature has employed the talents of another transition metal: manganese (Mn). Manganese works its magic in the oxygen-evolving complex (OEC) in the chloroplast where water is oxidized, releasing oxygen into the atmosphere and producing electrons and protons for the synthesis of organic molecules from CO_2.

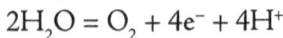

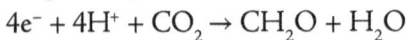

$$2H_2O = O_2 + 4e^- + 4H^+$$
$$4e^- + 4H^+ + CO_2 \rightarrow CH_2O + H_2O$$

As a result of recent advances arising from the use of a wide range of biochemical, biophysical, and molecular biological techniques, the precise role played by the manganese atoms in the OEC is now close to being understood.[54] At the heart of the OEC is a catalytic cluster containing four manganese atoms and one calcium atom. "Although the precise geometry of the Mn_4Ca cluster is not yet known precisely," Crichton says, it apparently "provides a basis for developing chemical mechanisms for water oxidation and dioxygen formation."[55]

And even though the precise details are not known of how the cluster of manganese atoms manages the miracle of oxidizing water and releasing oxygen into the atmosphere, one thing is clear: in nearly four billion years, no other mechanism for oxidizing water and releasing oxygen has evolved. The oxygen-producing system is essentially identical in all oxygenic photosynthetic organisms, from giant redwoods to blue-green algae (cyanobacteria). All oxygenic photosynthetic organisms use four manganese atoms and one calcium atom in the OEC. Indeed, the manganese-based mechanism appears to have arisen only once in the history of life on Earth,[56] a result consistent with the supposition that oxygen generation critically depends on the unique properties of the manganese atom.

Zinc

ZINC (ZN), although similar to some of the transition metals, is classed not as a transition metal but as a "Lewis acid" (a chemical species with an empty orbital that can accept an electron pair from a donor compound). There are about three grams of zinc in an adult human, most of which is intracellular.[57] Zinc is essential in all forms of life,[58] playing a role in more than 300 enzymes in each of the main fundamental classes: oxidoreductases, transferases, hydrolases, lysases, isomerases, and ligases. All of these roles in turn depend on particular properties of the zinc atom.[59]

While zinc is essential to all cells on Earth, one enzyme that uses zinc at its active site is specifically vital to terrestrial aerobic life—carbonic anhydrase (CA). This enzyme converts CO_2—the end product of oxidative metabolism—to bicarbonate (H_2CO_3) in the tissues, and converts bicarbonate to CO_2 in the lungs.

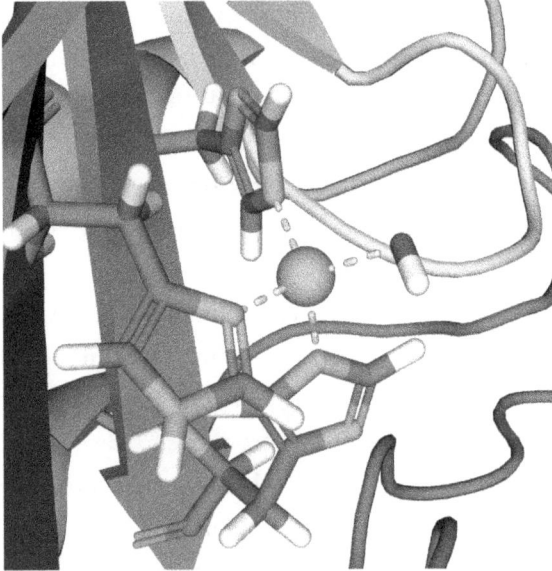

Figure 6.3. Rendering of human carbonic anhydrase II.

$$CO_2 + H_2O = H_2CO_3 \rightarrow CO_2 + H_2O$$

In the tissues—In the lungs

Carbonic anhydrase also aids in the regulation of fluid and pH balance and is involved in producing essential stomach acid. The enzyme also plays a role in vision. When it is defective, fluid can build up and cause glaucoma. The enzyme is one of the fastest known, catalyzing up to one million reactions per second.[60]

Might zinc be indispensable for this reaction? The evolutionary evidence suggests very strongly that this is likely. There are many types of carbonic anhydrase with no amino acid sequence similarities, suggesting that the enzyme has arisen many times independently in the history of life. But almost all have a similar active site, which contains zinc bonded to three nitrogen atoms in three histidine amino acids in the protein.

The discovery in 2008 that a marine unicellular organism can use cadmium (Cd) instead of zinc in certain circumstances suggests that zinc isn't uniquely required in carbonic anhydrase.[61] However, this discovery is not as radical as might appear, since cadmium is the homologue of zinc in period five of the periodic table. While the discovery does nuance our understanding, the result actually tends to confirm the claim that particular biological functions depend on particular metal ions. For one thing, the same basic active site was arrived at multiple times in evolution. For another thing, even though zinc and cadmium are exceedingly rare in some environments, no organism in billions of years of evolution has employed an alternative to zinc or cadmium. The implication is that only zinc or its close homologue can carry out this specific reaction with such extraordinary efficiency, a reaction that enables us to rid ourselves of some 100 million trillion molecules of CO_2 every breath.[62]

To briefly summarize before moving on to the final metal in this chapter: One metal, manganese, gives us oxygen. Two other metals, iron and copper, give us electron transport chains, proton pumping, and ATP. The oxidation of hydrocarbons in the mitochondria gives us H_2O and CO_2. And CO_2 requires another metal atom, zinc, if it is to be excreted from the body in the lungs. Together these provide powerful evidence for a stunning prior fitness in nature for aerobic life.

Molybdenum

ANOTHER TRANSITION metal necessary to life (and an essential nutrient in the human diet) is molybdenum (Mo). It occurs in four important enzymes in our bodies (sulfite oxidase, aldehyde oxidase, xanthine oxidase, and mitochondrial amidoxime reductase). People who don't get enough of it can suffer many adverse effects.[63]

Molybdenum also is involved in a vital activity upon which all life on Earth depends: nitrogen fixation. This process is carried out by the enzyme nitrogenase, which catalyzes the reduction of nitrogen to ammonia. Virtually all the nitrogen used by living things is initially captured by the work of this vital enzyme.

Nitrogenase fixes atmospheric nitrogen (N_2) by breaking the triple N bond which links the two nitrogen atoms together, reducing the nitrogen to ammonia NH_3. This is the compound through which the ni-

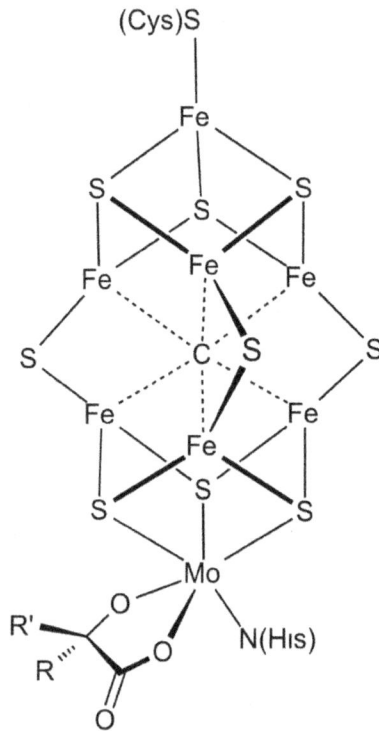

Figure 6.4. MoFe cofactor in the MoFe protein.

trogen atom is introduced into the organic domain. The reaction is very energy-dependent.[64]

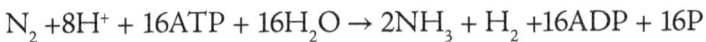

$$N_2 + 8H^+ + 16ATP + 16H_2O \rightarrow 2NH_3 + H_2 + 16ADP + 16P$$

Nitrogenase is composed of two protein subunits: the MoFe protein and Fe protein respectively. As the first name implies, the former contains molybdenum and iron, while the latter contains only iron. But there are alternative nitrogenases. In conditions when molybdenum levels are low, a nitrogenase which uses vanadium (V, a near homologue of Mo in the fourth period of the table of elements) in place of Mo. And when concentrations of both Mo and V are low, a nitrogenase in which Mo and V are replaced by Fe is synthesized. MoFe is highly complex and contains eight metal atoms—seven iron and (usually) one molybdenum—plus seven sulfur atoms.

Why, asks Crichton, does nature use such a precise complex assemblage of multiple elements in nitrogen fixation? Something simpler might have been expected, and yet what do we find? "Over more than a billion years, evolutionary pressures have retained this complex cofactor-based nitrogenase system," he writes, and "indeed, even the 'alternative' nitrogenases are thought to be minor variations on the cofactor with V or Fe replacing Mo."[65] He notes that all the metal atoms seem required and "to date no one has found a way to simplify the system."[66]

This is additional compelling evidence that certain metals—in this case, molybdenum, iron, and vanadium—are essential and uniquely fit to enable a highly specific reaction that is critical to all life on Earth. The absence of any alternative strongly suggests that the vital processes by which nitrogen enters the biosphere absolutely depend on the special characteristics of nitrogenase and its complement of metal atoms.

Metals Most Precious

WHEN WE speak of precious metals, we usually mean elements such as gold and silver. But it is upon several relatively common and affordable metal elements that life most depends. The miracle of the cell hinges on the fact that these various metal atoms possess differing and very specific physical and chemical properties. Although not quite as diverse as the properties of the nonmetals, they are sufficiently different one from another—in ionic dimension, redox potentials, coordination chemistry, and so on—to confer on each of these elements unique properties which have been exploited for highly specific biochemical purposes. Those purposes include: activating ATP; forming electron-conducting wires in chloroplasts, mitochondria, and bacterial cell membranes; carrying charge across cell membranes at great speed; and acting as vital cofactors in multiple different enzymes.

The empirical evidence speaks for itself. Particular metal atoms have performed the same cellular function over billions of years. Given the astonishingly diverse environments exploited, for example by various microbial extremophiles[67]—temperatures of 120°C, acidic environments less than pH 1, alkaline environments at pH 10.5, hypertonic saline solutions ten times as salty as sea water—the fact that the roles of the metals are the same in all these organisms speaks volumes for the special fitness of these atoms for highly specific cellular functions. Without the metals there might be some very primitive type of carbon-based life, but anything remotely as rich and complex as now exists would be impossible.

All life forms on Earth use transition metals in electron transport chains (ETCs) that, by all appearances, are indispensable to life. In four billion years of evolutionary experimentation, since the first blue-green algae began its fateful manufacture in a vast number of diverse lineages, the same choice of transition metals have been made again and again to build ETCs. The best explanation is that no other elements, nor any types of organic compounds, can replace the transition metals for this task.

Of particular interest to us is the special fitness of transition metals for the unique functions on which we advanced aerobic organisms depend. These include the fitness of magnesium for light-capture in photosynthesis, the fitness of iron and copper for the transport and activation of oxygen, and the fitness of zinc for the function of carbonic anhydrase (and consequently for the excretion of CO_2). And of course, the highly mobile atoms sodium and potassium are uniquely fit for generating membrane potentials, which enable nerve impulse transmission. Without them there would be no central nervous system and certainly no carbon-based intelligent beings like humans.

Our oxygen-hungry lifestyle absolutely depends on the properties of metal atoms in the center of the periodic table. In the end, Williams was right when he concluded that metals are just as critical to biology as DNA or proteins.[68]

7. The Matrix

Water is the most important liquid for our existence and plays an essential role in physics, chemistry, biology and geoscience. What makes water unique is not only its importance but also the anomalous behaviour of many of its macroscopic properties.... If water would not behave in this unusual way it is most questionable if life could have developed on planet Earth.
—Anders Nilsson and Lars G. M. Pettersson[1]

T HE TINY LEUKOCYTE CELL CHASING A BACTERIUM IN THE VIDEO mentioned in Chapter 1 makes compelling viewing.[2] The leukocyte senses the chemical "aroma" of the bacterium. It crawls in a specific direction along a chemical gradient, following the smell of the bacterium. And it does this while continuously making transient selective adhesions to the substratum. These skills are remarkable any time, but we find by far their most spectacular application in the miracle of embryology.

In the embryo it is not just one cell moving towards a specific target but millions of cells, each moving towards specific targets in an ever-changing kaleidoscope of different embryonic cells and chemical signals, with each cell obeying a strictly choreographed program, a program directing the timing of gene expression and a unique succession of changes in cell shape and cell surface proteins and adhesive properties in different cells in different regions of the embryo.

While crawling through the embryo and sticking to their assigned partners at precise moments and in precise places in the developing embryo, the cells continually pick up chemical and physical signals from the mass of surrounding cells, signals that direct all manner of molecular

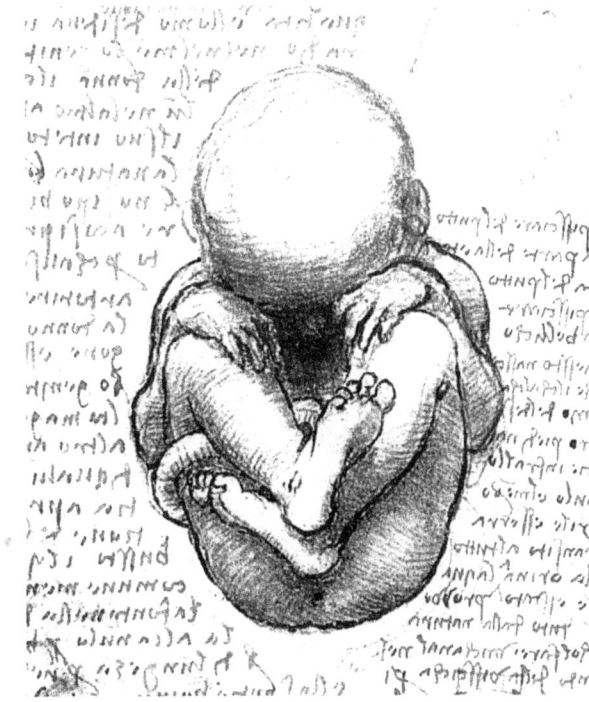

Figure 7.1. Leonardo da Vinci's illustration of a fetus in the womb.

and genetic activities within the individual cells. These involve precise-ly timed expression of particular sets of genes and exquisitely ordered movements of actin filaments, molecular motors, and microtubules, which act together to generate continual changes in the cells' architec-ture. In one place and time an embryonic cell may turn into a neuron, in another into a red blood cell, in another into a leukocyte, in another into a photoreceptor.

It is hard to envisage any material phenomenon more complex than the development of the embryo, involving as it does the integra-tion of such a myriad of precisely coordinated events. The totality is be-yond grasp. It involves a seeming infinity of subtle and exquisitely cho-reographed changes in the architecture and shape of cells in different parts of the embryo; precise spatial and temporal ordering of myriads of changes in cell surface properties in different parts of the embryo; numerous precisely ordered cell divisions, particularly in the earlier stages of embryogenesis; a vast diversity of emergent cell movements and cellular morphologies; the setting up of all manner of diffusion gradients; and the programming of numerous cell types to read their exact position in these gradients at precisely predetermined times.

The development of the embryo with its staggering panoply of continuously morphing cells—each finding its unique way through the seething and dynamic yet highly ordered embryonic web of cellular matter, touching and feeling its neighbors in search of spatial and temporal clues and obediently changing its own chemical, genetic, and physical state in response—is by far the most complex phenomenon on Earth, far more complex by many orders of magnitude than the assembly of the most complex human artifact ever built.

Indeed, the developing embryo is a phenomenon far beyond anything in the realm of our ordinary, or extraordinary, experience. The unimaginable immensity of spatial and temporal molecular clues and molecular and genetic responses exploited by this innumerable host of nanobots navigating the embryonic ocean is far greater than all the maps, charts, and devices used by all the mariners who ever navigated the oceans of Earth. No human machine built to date nor any on the drawing board of even the most ambitious and farsighted gurus of nanotechnology is remotely as complex as a developing embryo.

One key to the miracle is another ensemble of fitness in nature, this one arising from the unique properties of that most familiar yet most remarkable of liquids—water. As Albert Szent-Györgyi famously commented, "Water is in any case the central substance of living nature. It is the cradle of life, the mother of life and its medium. It is our mater and matrix."[3]

Early chapters touched on some of the unique properties of water essential to cellular function. Several of these and others are described in more detail below.[4]

Viscosity

ONE PROPERTY of water not touched on in this book thus far, which has a critical bearing on the function of all carbon-based cells but particularly on the function of the large complex cells in advanced multicellular organisms like ourselves (and embryos) is its viscosity. Water's viscosity determines two important parameters. One is the diffusion rate of solutes such as oxygen and nutrients in aqueous solutions. The second, *viscous drag*, is the resistance experienced in moving an object in a liquid (think of moving a spoon through honey). Diffusion rates are inversely proportional to the viscosity of a liquid (as indicated by the Stokes-Einstein equation, $D = k/m$, where D is rate of diffusion, k is a constant and m is viscosity[5]) while viscous drag is directly proportional to viscosity.[6] As we shall see, if the viscosity of water was not very close to what it is, there would be no embryos, no multicellular organisms, and no cells capable of the crawling and morphing talents of the leukocyte and its embryonic cousins.

Diffusion

NOT ALL cells can crawl or change shape like a leukocyte or the cells of an embryo. Many cells, including plant and fungal cells, are encased in a rigid cell wall and are unable to crawl or change shape. Bacterial cells can move using a flagellum, but they cannot crawl or adopt an endless series of different morphologies as the leukocyte can. Only cells bounded by the pliable and remarkably deformable bilayer lipid cell membrane can crawl, change shape, and use these two skills to build the embryo.

The absence of a rigid cell wall is not the only factor that precludes the escaping bacterium from being able to crawl and morph like the leukocyte. Cell motility and morphing also depend on a suite of specialized cytoplasmic molecular motors and other macromolecular components. Their extraordinary complexity is described in many major texts.[7] Most bacterial cells are far too small to contain this suite of cytoskeletal components. An *E. coli* bacterial cell, for instance, may contain a mere two million proteins,[8] compared to many human cells containing several billion protein molecules,[9] several million actin monomers (which polymerize to make actin filaments), and about a million myosin motor molecules (providing the contractile force involved in crawling).[10]

The maximum possible diameter of cells is constrained by a fundamental physical parameter: the rate of diffusion of molecules in water. Diffusion is very rapid and effective over short distances but increasingly slow and inefficient over long distances. More formally, diffusion time increases with the square of the diffusion distance. Thus, as Knut Schmidt-Nielsen explains, if one were to impose a stepwise increase in oxygen at a given point, it would diffuse one micron in one ten-thousandth of a second. But for the oxygen to travel ten times as far would take one hundred times as long. Thus an average diffusion distance of ten microns (the distance across some cells) takes one hundredth of a second; one millimeter takes one hundred seconds; ten millimeters, about three hours; and one meter, about three years.[11]

Since diffusion time increases with the square of diffusion distance, and volumes increase with the cube of a sphere's diameter, it follows that if diffusion rates in water had been slower than they are—that is, as slow as they are in many other fluids (see below)—utilizing oxygen and nutrients at the same rate as our own body cells or the cells of the embryo use them would require decreasing the sizes of the cells substantially. For example, if diffusion rates were ten times less than they are, then maintaining the same oxygen consumption would require the volume of a roughly spherical cell to shrink by a thousand fold. If diffusion rates were one hundred times less than they are, then the same cell would need to shrink by a million fold to maintain the same oxygen consumption—in both cases, almost certainly too small to contain the complex cytoskeletal components to enable crawling and the other abilities cells need to create an embryo.

Also, because the surface area of such hypothetical mini-cells would be one hundred to ten thousand times less, the total number and diversity of cell-surface adhesion molecules also would be drastically reduced. Such tiny cells would hardly be able to put out the complex arrays of micro-protrusions or filipods (mentioned in Chapter 4) that detect chemical signals in the environment and initiate cell-cell and cell-matrix adhesion.[12] These abilities are crucial for the pathfinder or pioneer neurons in the developing nervous system.[13]

In sum, the existence of cells large enough to crawl and morph depends on the diffusion rates of oxygen and other solutes in water being not significantly less than what they are. If these diffusion rates were appreciably lower, active aerobic cells large enough to contain the molecular machinery for crawling and morphing could never have emerged.

Viscous Drag

BECAUSE OF diffusional constraints, the aerobic cells in complex multicellular organisms, including embryos, require a set of narrow tubes—capillaries about five microns across—that permeate all the tissues and organs of the organism. These carry blood enriched in oxygen, which supplies via diffusion sufficient oxygen to satisfy the energetic demands of these organisms. Mammalian tissues, for example, are permeated by 1,000 capillaries per square micron[14] so that most capillaries are about forty microns apart and most tissue cells are between one to three cell widths from a capillary.[15]

The design of the capillary bed is constrained by another physical parameter arising from viscosity—viscous drag. The pressure (P) required to pump a fluid through a pipe rises with the viscosity of the fluid (m),[16] because the greater the viscosity the greater the viscous drag. Think of the difficulty of drawing honey up a straw compared with a far less viscous fluid such as water.

If blood's viscosity (largely determined by water's viscosity) were increased two or three times, the pressure needed to pump blood through the capillary bed would be too great. As things are, the head of pressure at the arterial end of a human capillary is thirty-five mm Hg, which is considerable, about one-third that of the systolic pressure in the aorta. This relatively high pressure is necessary to force blood through the capillaries. This would have to be increased proportionately if water's vicosity were higher. But then the thin endothelial lining of the capillaries (necessary if nutrients and oxygen are to diffuse unimpeded into the tissues) would be subject to greater pressures and the threat of rupture.

From these considerations it is clear that the functioning of the circulatory system and particularly the capillary bed depends on two different physical parameters both being very close to what they are: the diffusion rates of oxygen and nutrients in water, and the viscous drag of water. The high diffusion rates enable the transfer of oxygen and nutrients from the capillaries in sufficient quantities to supply the energy needs of the cells, but only because water's low viscous drag enables the perfusion of the capillary bed at a rate sufficient to satisfy those needs.

If viscosity were significantly higher, trying to compensate for that by increasing the number and size of the capillaries would take up too much volume. The organism or embryo would be reduced to a bag of fluid, leaving no room, as Stephen Vogel put it, for "guts or gonads."[17]

On the other hand, if the viscosity were much lower, then diffusion rates would be greater and viscous drag less, but the delicate cytoarchitecture inside the cell would be subject to more intense Brownian bombardment, since particle mobility in a fluid is inversely related to viscosity.[18] Thus, things would be far less stable. The half-lives of the cell's key macromolecules would also be decreased and the energetic burden of maintaining cellular homeostasis increased. It is highly doubtful if anything remotely resembling a living cell would be feasible if the viscosity of water approached, for instance, that of a gas. The low viscosity of gases is the reason most authors reject them as suitable media to instantiate a chemical living system like the carbon-based cell.[19] Gases are also far too volatile and labile to be considered seriously as candidates for the chemical matrix of life, as Arthur Needham comments:

> Only systems based on a fluid medium could display the properties which we should accept as Life. Gaseous systems are too volatile and lack the powers of spontaneously segregating sub-systems... In gases... the tendency towards a uniform distribution of energy is very rapid but in liquids is slow enough for local differences to be maintained... and

for steady-states to be set up. All gases mix freely with any other, but not all liquids. Some form discontinuities or interfaces, with solid-state properties, where they meet another liquid, and complex polyphasic systems are readily formed which further… increase the potentialities for steady-state perpetuation.[20]

Needham notes that clouds are a rare exception to the rule that segregating sub-systems are unusual in a gas. However, the very transience of cloud patterns graphically illustrates the unsuitability of gas as a medium for the support of stable segregating sub-systems.

It is not possible to determine precisely how narrow is the range of viscosities and diffusion rates compatible with a functioning circulatory system and aerobic cells large enough to possess the crawling and shape-morphing abilities of a leukocyte (or a motile embryonic cell). But all the evidence suggests that it must be very close to what it is, within a range of about 0.5 mP-s to 3 mP-s.

That the fitness of the viscosity of water must fall within such a narrow range highlights just how fine-tuned is the natural order for life. The viscosity of common substances varies greatly.[21] Measured in millipascals-seconds, the viscosity of air is 0.017, water 1.0, olive oil 84, glycerin 1420 and honey 10,000.[22] The total range of viscosities of substances on our planet is more than twenty-seven orders of magnitude, from the viscosity of air to the viscosity of crustal rocks.[23] Thus, the range of life-friendly viscosities is a tiny, vital band within the inconceivably vast range of viscosities in nature.

In sum, it is ultimately water that permits cells in a multicellular organism or developing embryo to be adequately supplied with oxygen and nutrients via a circulatory system. It is also water that permits cells in multicellular organisms to grow large enough so that they can crawl and morph. Only then can cells in the embryo perform that greatest of miracles—the directed assembly during development of the trillions of embryonic cells into the mature form of the organism.

The Universal Solvent

ANY FLUID serving as the cell matrix must also be a good solvent, able to carry in solution a vast range of ions and biochemicals, including oxygen and various nutrients necessary for cells to function. Water wonderfully fits the bill.[24]

As a solvent, water is without peer. "Liquid water is such a good solvent, in fact, that it is almost impossible to find naturally occurring pure samples and even producing it in the rarefied environment of the laboratory is difficult," writes Alok Jha. "Almost every known chemical compound will dissolve in water to a small (but detectable) extent. Related to this, because it will interact with everything, over long periods of time water is also one of the most reactive and corrosive chemicals we know."[25] The sentiment is echoed by the late Felix Franks, one of the leading authorities on the properties of water: "The almost universal solvent action of liquid water" makes "its rigorous purification extremely difficult. Nearly all known chemicals dissolve in water to a slight, but detectable extent."[26]

Water is indeed the alkahest, the universal solvent the alchemists sought. No other liquid comes close. As Lawrence Henderson commented, "As a solvent there is literally nothing to compare with water... In the first place the solubility in water of acids, bases and salts, the most familiar classes of inorganic substances, is almost universal."[27]

Virtually all organic substances that carry either an ionic charge or contain polar regions—which include the great majority of all organic compounds in the cell and biological fluids—dissolve readily in water. This includes proteins and other big molecules, provided they have polar or ionic regions on their surfaces.[28] (See Chapter 4.)

The Hydrophobic Force

AS WE also saw in Chapter 4, there is one exception to the solvation powers of water, which is worth briefly recalling. It is common knowledge that oil and water don't mix. If you pour oil on water, it doesn't dissolve. Instead it forms a separate layer on the surface. The reason is that oils

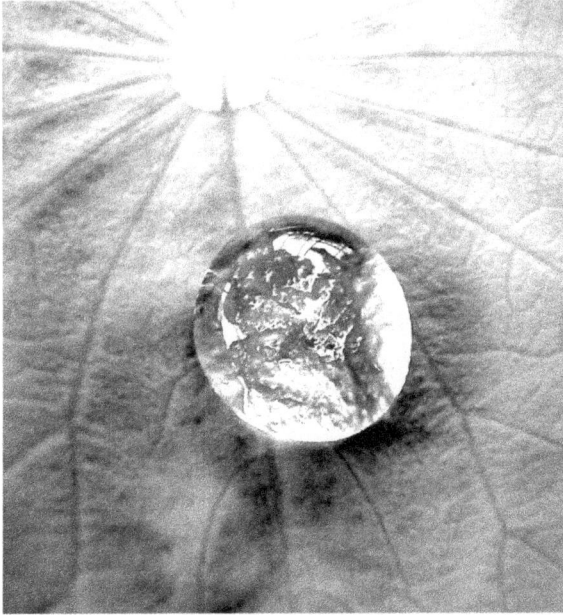

Figure 7.2. A water drop on a leaf shows the power of the hydrophobic force.

consist of long hydrocarbon chains in which C-H bonds are nonpolar (the electrons are equally distributed over the carbon and hydrogen atoms in the chain), so there are no positively or negatively charged regions along the chain. This lack of charged regions precludes the formation of hydration shells, which are involved in rendering a substance soluble in water.

The hydrophobic effect can be observed in the beading of water on nonpolar surfaces, such as waxy leaves after a rain shower. The water molecules are unable to form hydrogen bonds with the waxy hydrophobic surface and are forced to clump together into beads away from the surface.

As we saw in Chapter 4, both the assembly of cell membranes and the folding of proteins depend on hydrophobic forces. Water's inability to dissolve hydrophobic compounds is no defect but a vital element of water's fitness for the carbon-based cell.

The Active Matrix

As SCIENCE has progressed and the properties of bio-matter have become better understood, new elements of the fitness of water for life have been revealed. Recent research has shown, for example, that water is a far more active player in cellular physiology than was previously believed, thoroughly supporting Albert Szent-Györgyi's celebrated claim that "life is water dancing to the tune of solids."[29]

Philip Ball concurs: "It has become increasingly clear over the past two decades or so that water is not simply 'life's solvent' but is indeed a matrix more akin to the one Paracelsus envisaged: a substance that actively engages and interacts with biomolecules in complex, subtle, and essential ways."[30] Ball gives as an example DNA, noting that its double helical structure "relies on a subtle balance of energy contributions present in aqueous solution." If water weren't there to "screen the electrostatic repulsions between phosphate groups," the double helix would cease to be viable, as evidenced by the fact that "DNA undergoes conformational transitions, and even loses its double helix, in some apolar solvents."[31]

In concluding his article, Ball sums up the developing view:

Water plays a wide variety of roles in biochemical processes. It maintains macromolecular structure and mediates molecular recognition, it activates and modulates protein dynamics, it provides a switchable communication channel across membranes and between the inside and outside of proteins. Many of these properties do seem to depend, to a greater or lesser degree, on the "special" attributes of the H_2O molecule, in particular its ability to engage in directional, weak bonding in a way that allows for reorientation and reconfiguration of discrete and identifiable three-dimensional structures. Thus, although it seems entirely likely that *some* of water's functions in biology are those of a generic polar solvent rather than being unique to water itself, it is very hard to imagine any other solvent that could fulfill all of its roles—or even all of those that help to distinguish a generic polypeptide chain from a fully functioning protein.[32]

The increasing evidence and awareness that water may be an active player in cellular physiology is described in the text *Water and the Cell,* which defends the notion in the preface: "Water within cells is to a major extent ordered differently than bulk water, and functions not as an inert solvent, but as an active player... Understanding water order in biological systems is key to an understanding of life processes."[33] Some of the chapter titles emphasize just how important water's active role may turn out to be for cellular physiology: "Information Exchange within Cellular Water," "Biology's Unique Phase Transition Drives Cell Function," "Some Properties of Interfacial Water," and "Biological Significance of Active Oxygen Dependent Processes in Aqueous System."

Considering the fantastic number of ways water is fit for life on Earth and how more examples keep coming to light as scientific knowledge advances, it is likely there are many more to be discovered. Despite its importance and the massive research effort[34] devoted to understanding it, the structure of intracellular water is still more mysterious than the geophysics of Mars or the chemistry of hydrothermal vents.

Proton Wires

ONE INTRIGUING element of fitness for bioenergetics and proton pumping arises directly out of water's hydrogen-bonded network,[35] which provides so-called "proton wires" consisting of long chains of linked water molecules for moving protons (H ions) around in the cell and across the inner mitochondrial membrane.

While, as Alok Jha points out, other charged particles involved in cellular functions have to move themselves physically from one place to another, "protons can pass their energy along a hydrogen-bonded water wire without moving themselves at all, thanks to the so called Grotthuss mechanism." A proton attaches to one end of the wire, he explains, and in a split second, "each of the hydrogen bonds further along the length of the wire spin around in sequence so that a proton drops off the water molecule at the other end of the wire. The initial proton has not moved any further than the starting end of the wire but its charge and energy have been 'conducted' along the wire's length."[36]

Biophysicist Harold Morowitz discusses the unique fitness of these water wires for bioenergetics. "The past few years have witnessed the developing study of a newly understood property of water [proton conductance] that appears to be almost unique to that substance, is a key element in biological-energy transfer, and was almost certainly of importance in the origin of life," he writes. "The more we learn the more impressed some of us become with nature's fitness in a very precise sense."[37]

Nick Lane, author of *The Vital Question*, also sees proton conductance as playing an essential role in the origin of life, driving the formation of organic compounds.[38] If Lane is correct and proton flows provided essential energy for the synthesis of organic compounds in the earliest protocells,[39] then water would indeed be the cradle and mother of life, as Szent-Györgyi claimed more than fifty years ago.

The Prime Coincidence

THERE IS one further very critical element of water's fitness for cellular life to consider, touched on in a previous chapter: the temperature range in which living things thrive on Earth, and which is fit for the chemistry of life, is almost exactly the same temperature range in which water is a liquid.

The currently established upper temperature limit for metazoan organisms is about 50°C.[40] This limit applies even to animals adapted to the hot waters of hydrothermal vents, which can only survive short periods at temperatures above 45°C.[41] One exception is a species of desert ant, which can survive for short periods at temperatures of 55°C.[42] Current record-holders for the most thermo-tolerant microorganisms are hyper-thermophilic species of bacteria which can survive in temperatures of as much as 120°C and are found in hot springs throughout the world, such as those in Yellowstone Park in the United States and in water close to oceanic hydrothermal vents. (In the deep ocean, water temperature can climb over 100°C because the pressure is much higher than atmospheric pressure at sea level.) The current record-holder is a methanogen discovered in the Indian Ocean, in black smoker fluid of the Kairei hydrothermal field. It survives and reproduces at 122°C.[43]

As mentioned previously, the lower limit for life is hard to determine exactly because water freezes at 0°C. However, many organisms protect themselves against freezing by cryoprotectants, and microbial metabolism has been reported down to −20°C.[44] A midge in the Himalayas survives in temperatures of −18°C.[45] However, at temperatures much below −20°C, no matter what cryoprotectants are used, freezing inevitably occurs. Even if ice crystal formation is avoided, intracellular water eventually undergoes a glass transition—vitrification. This effectively stops all metabolic processes.[46]

Consequently, a temperature of about −20° marks the lower recorded limit of life, or at least of active metabolism on Earth. This limit is imposed by the properties of the medium in which all life processes occur—water[47]—and thus might seem a defect in water's fitness for life. However, if metabolism could be maintained at colder temperatures in something other than water, in a matrix that was a liquid at colder temperatures (ammonia comes to mind, which is a liquid between −78°C and −33°C), it would be glacially slow.[48] In fact, life at −40°C would be sixty-four times slower than life at −20°C. Or, more properly, the rates of chemical reaction would be sixty-four times slower.

This means that at whatever constraints the rate of biochemical processes impose on biological evolution in our actual world, that rate would arguably be as much as sixty-four times slower in such a hypothetical cold world. It took more than 300-million years to get from multicellular jellyfish (Haootia) to the first mammals. Multiply that by sixty-four and we're looking at around 20 billion years for evolution to follow that path. By then the sun would long have burned out, rendering Earth uninhabitable.

A world capable of evolving creatures like ourselves, then, would appear to require a matrix with a liquid range similar to water's, or at least not one where life must exist at substantially colder temperatures.

What about the possibility of carbon-based life at temperatures much above 100°C? It is likely that the currently highest known temperature in which bacteria can survive, 122°C, is close to the maximum possible for carbon-based life on Earth. And it must be presumed that this upper limit is imposed because, as discussed previously, the organic compounds used in the cell become increasingly unstable as temperatures rise above 100°C. In fact, whatever theoretical reasons one might adduce for the upper temperature range of life (e.g., the instability of covalent and particularly weak bonds), nature has in effect already indicated the upper temperature limit empirically by the absence of any organisms on Earth able to withstand temperatures much above 100°C. After four billion years of experimentation, nature has provided clear empirical evidence that the fecund chemistry of carbon is of little use in constructing biological systems at temperatures beyond this.[49]

The weak bonds, being ten to twenty times weaker than covalent bonds, are even more susceptible to thermal disruption as temperatures rise above the ambient range. As I noted in *Nature's Destiny*, the relative weakness of the weak bonds compared to the strong covalent bonds is apparent in cooking. "The disruption of weak bonds occurs in two very familiar processes in the kitchen—in the heating and beating of egg white, both of which causes the egg white to whiten and coagulate"[50] and also in the slow cooking of meat between 85° and 90°C. The gentle heating causes the familiar softening of the meat. Neither the beating of egg whites nor the slow cooking of meat under 90°C breaks the covalent bonds in the egg white or the meat. The softening of the meat and the coagulation of the egg white involve the breaking of the weak bonds which hold proteins in their native 3-D shapes, causing their unraveling and denaturation, which occurs at temperatures considerably below those required to break covalent bonds (beyond 100°C for most covalent bonds). Because covalent bonds are far more robust, neither beating nor gentle heating has any significant effect on them and most remain intact. The extreme sensitivity of the weak bonds to high temperatures is another reason why few organisms can survive at temperatures above 100°C.[51]

Although a temperature range of –20°C to 122°C (a range of 142°C) appears from our mundane perspective to be considerable, as pointed out in Chapter 2 such a range is an unimaginably tiny fraction of the total range of all temperatures in the cosmos. Temperatures in the cosmos range from 10^{32}°C (10 followed by thirty-one zeros), which was the temperature of the universe shortly after the Big Bang, to very close to absolute zero, or –273.15°C. The temperature inside some of the hottest stars is several thousand million degrees.[52] Even inside our own Sun, which is not a particularly hot star, the temperature is on the order of fifteen million degrees,[53] and its surface temperature is just below 6,000°C.[54] So, out of the enormous range of temperatures in the cosmos, there is only one tiny temperature band, about one 10^{-29} of the total range, where water is a liquid. Within this tiny temperature band, the energy levels of the covalent bonds of the organic domain can be manipulated by living systems; the weak bonds can be used for stabilizing the 3-D forms of complex molecules; and water, the only compound known to possess the many other properties essential to serve as the matrix of life, exists in the liquid state.

This is little short of a miracle. If this coincidence did not hold, water would not be fit to form the matrix of the cell. All the myriad other elements of fitness of this unique fluid would be to no avail. Almost certainly there would be no carbon-based life in the cosmos. Or to put that matter more positively, it is surely an awesome coincidence, indicative of the profound fitness of water and, by extension, of nature for carbon-based life, that the optimum temperature range for the complex atomic and molecular manipulations essential for life is precisely the temperature range in which water, the ideal matrix in so many other ways, exists as a liquid at ambient conditions on Earth.

Unrivaled

THE EVIDENCE is overwhelming: water possesses an ensemble of unique properties which uniquely equip it to function as the matrix of the carbon-based cell. There is hardly an author conversant with the facts who would contest Henderson's verdict that no substance can rival the fitness of water as the *milieu intérieur* of carbon-based life.[55] And discoveries since Henderson's time have merely deepened and broadened the evidence for that conviction. Kevin Plaxco and Michael Gross recently reviewed the fitness of water in the introductory chapter of *Astrobiology* and concluded that water's "ability to form the basis of biochemistry may well be unique... no other liquid has even a fraction of the favorable attributes of water."[56]

Finally, water also has many unique properties essential for creatures of our physiological design. The high diffusion rate of solutes in water makes possible a circulatory system to supply our body cells with oxygen and nutrients. It also makes possible cells sufficiently large to contain the necessary cytoplasmic machinery for cellular motility and shape changing, essential for so many cellular functions, including embryo assembly. Again, our warm bloodedness depends critically on the thermal properties of water—its high heat capacity and high latent heat of evaporation. Water also plays a vital role in the transport of carbon dioxide from the tissues for excretion in the lungs. In short, in the properties of water, nature appears to have had not only life in mind, but creatures like ourselves.

8. THE PRIMAL BLUEPRINT

An honest man, armed with all the knowledge available
to us now, could only state that in some sense, the origin
of life appears at the moment to be almost a miracle, so
many are the conditions which would have to be satisfied
to get it going.

—FRANCIS CRICK, *LIFE ITSELF* (1981)[1]

THERE IS SOMETHING EVOCATIVE AND ENDLESSLY FASCINATING
about the idea of a message sent to Earth from space by an advanced
extra-terrestrial civilization, revealing information about our own origin, existence, and place in nature. It is the stuff of many popular works,
from Carl Sagan's *Contact* and Stanley Kubrick's *2001: A Space Odyssey*
to Erich von Däniken's book *Chariots of the Gods*.

Although not the work of aliens, a very special chemical message
regarding the place of carbon-based life in the universe did arrive on
Earth on the night of September 28, 1969.[2] On that night, the sky over
the small southeastern Australian town of Murchison was lit up by an
exploding meteorite, which scattered fragments of rock over the nearby
countryside. Subsequent chemical analyses of the meteorite's fragments
proved, for the first time, that at least some of the organic building blocks
of life are being constantly synthesized in space and exist in vast quantities throughout the cosmos.[3] Moreover, very recent analyses suggest the
total number of different carbon compounds embedded in the meteorite
may number in the tens of thousands and perhaps even millions.[4]

The Murchison meteorite revealed that the cosmos is seeded with
a vast inventory of organic chemicals, including amino acids[5] and
nucleobases,[6] the starting points in the assembly of the two main poly-

mers—proteins and nucleic acids—in carbon-based organisms. Just how many of the basic organic building blocks of life might have been produced abiotically in space and brought to Earth in meteorites (or synthesized in the primeval ocean) is not clear. But more and more are being identified in meteorites and synthesized in the lab in simulated prebiotic conditions.[7]

More recent spectroscopic analyses of interstellar gas have revealed that the cosmos is seeded not just with some of the basic monomers of life, but also with far more complex carbon compounds. One class of complex carbon compound—the polycyclic aromatic hydrocarbons (PAHs)—is abundant throughout the cosmos, and some may contain up to one hundred carbon atoms.[8] Some contain nitrogen (PANHs), forming heterocyclic compounds that have a chemical structure similar to the heterocyclic compounds used in living things, such as the nucleotide bases.[9]

It is not hard to see a parallel between Friedrich Wöhler's landmark synthesis of urea in 1828 and the message the Murchison meteorite delivered. Wöhler undercut the need for a vital force in the cell for manufacturing the basic organic constituents of life, but their synthesis still required a cell or a chemist. The Murchison meteorite and subsequent study of other meteorites and spectroscopic analysis of interstellar space imply that the universe is replete with organics and that when the inventory is more fully known, it very well may contain many more of the basic building blocks of life. These have been synthesized, to paraphrase Wöhler, not only without a kidney, but even without a cell or a chemist. And these discoveries have, to a large extent, validated a widespread belief of origin-of-life researchers in the twentieth century, since Stanley Miller's ground-breaking experiment in 1953, that the basic ingredients of life can be synthesized abiotically in nature.

The Miller-Urey experiment involved sending a spark through an atmosphere thought to mimic the atmosphere of the primeval Earth (water vapor, methane, ammonia, and hydrogen). The result was a complex chemical mix that contained glycine, alanine, and aspartic acid, three of the amino acids used in building proteins in modern organisms.[10] Recent studies of his archived material from his original experiments subjected to more advanced analytical techniques show that ten of the twenty biologically important amino acids—lysine, alanine, serine, threonine, aspartic acid, valine, glutamic acid, methionine, isoleucine, and leucine—were present in Miller's flasks.[11]

Of course the three amino acids reported by Miller, and even the ten in the later accounting, are a long way from the inventory of monomers needed to assemble a living cell. Much less does it explain how the inventory of monomers were assembled to form the first cell. Nonetheless, Miller-type experiments and evidence from the analysis of meteorites such as the Murchison meteorite do show that at least some of the key building blocks of life can be, and indeed are, synthesized abiotically, and may be common throughout interstellar space. This is no small discovery. For they show that at least the first step towards the cell requires no more than "ordinary chemistry" and suggests that the subsequent steps to the cell might also be explicable in terms of the known laws of chemistry and physics—that nature might be sufficient.

Thus we see that the message brought to Earth that fateful September night, written in the chemistry of a falling star, is highly significant. As well as supporting the claim that life's emergence might have been the result of entirely natural mechanisms, it also supports the thesis central to this book and the whole *Privileged Species* series: Carbon-based life as it exists on Earth is no contingent afterthought of nature, no artifactual accident, but an inherent part of the natural order. It is an inherent part of nature's grand design from the moment of creation.

Cosmic Abundance

THERE IS further evidence that life may be no contingent cosmic afterthought but an end programmed into the order of things from the beginning. The atoms carbon, oxygen, and nitrogen were among the first atoms synthesized in the stars. These joined hydrogen, already in existence, to make up the universe of organic chemicals, the substances upon

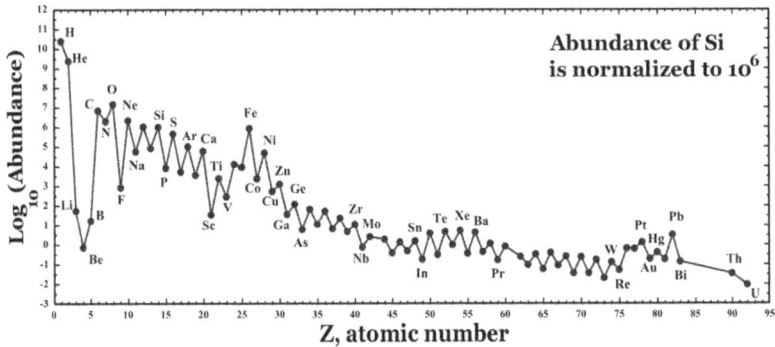

Figure 8.1. Cosmic abundance of elements.

which Earth's whole carbon-based biosphere is built. (When Lawrence Henderson wrote his classic *Fitness*, the source of carbon, oxygen, nitrogen, and various other elements heavier than hydrogen and helium was a mystery, since the nuclear synthesis of the atoms of the periodic table in the interior of stars was only elucidated by Fred Hoyle and others in the mid-twentieth century.) These four atoms are also, along with helium, commonest of all atoms in the universe.[12]

Even a cursory observation of the cosmic abundance of the elements reveals an obvious correspondence between the cosmic pattern generated in the heart of the stars and life on Earth, between the cosmic order and the order of life, between man and cosmos. The elements hydrogen (H), carbon (C), oxygen (O), and nitrogen (N), the core atoms that combine to form the molecules of organic chemistry that compose 96 percent of the human body, are respectively the first, third, fourth, and fifth most abundant elements in the cosmos. And their order of abundance curiously corresponds to their abundance in the human body.[13] Two of the three most abundant elements, hydrogen and oxygen, make up water (H_2O), the matrix of life, which forms more than 60 percent of the mass of the human body. And another key molecule, carbon dioxide (CO_2)—the ideal carrier of the carbon atom to all life on Earth—is formed from the third and fourth most abundant atoms: oxygen and carbon.

Other prominent constituents of life are also among the most abundant of elements: magnesium, sodium, calcium, iron, phosphorus, potassium, and sulfur. The overall picture conveys the powerful impression that stellar nuclear synthesis—the atom-building process in the stars—was set up from the beginning to serve the end (the purpose) of life on Earth.

It is important to stress that the selection of the atoms which enable the biochemistry of life is not because of their cosmic abundance, but because they possess the right chemical and physical properties to serve a vast number of highly specific physiological and biochemical functions in the cell.[14] Carbon, for instance, possesses the right properties to build a vast diversity of chemical compounds, and this would be true even if it were not the fourth most abundant element in the universe. No other atom possesses the same fitness. In other words, the laws which determine the cosmic abundance of carbon and the other atoms of life are quite distinct from the laws that determine their fitness for life. So here is a genuine coincidence indicative of a deep biocentricity in the cosmic order: the great majority of the most abundant atoms are the most fit for life.

The Elusive Path

DESPITE THE message of the Murchison meteorite—that the cosmos is seeded with the atoms of life and even with many of the core organic molecules, including amino acids and nucleobases—just how or where the transition from soup to cell occurred is an abiding mystery, among the greatest unsolved problems in science.

We do know that banded stromatolite formations generated by mats of bacterial cells very similar to modern blue-green algae first appear in the fossil record about 3.5 billion years ago,[15] and there is some fossil and isotopic evidence which implies life might have originated as early as 3.7 billion years ago.[16] So we know life has graced our planet for billions of years. But we know virtually nothing about how it originated.

The depth of the mystery is compounded by the fact that we do know, as the previous chapters in this book show, a great deal about the chemical basis of life. We have known since the early nineteenth century that the building blocks of the cell are perfectly natural chemical forms, determined by natural law. After the Murchison discovery and recent astrophysical studies, we know that at least some of these building blocks occur in vast quantities throughout the cosmos. We also know, in astounding detail, the atomic structures and molecular behavior of the key macromolecular constituents of the cell—such as DNA and RNA, proteins, and lipid membranes. We also have known since the 1960s the basic design of the cell, the meaning of the genetic code, and how information flows from the DNA into proteins. More recently, we have uncovered undreamt-of depths of complexity in the genome, including a mushrooming inventory of tiny regulatory RNAs.[17]

The size of standard textbooks like *Molecular Biology of the Cell*[18] is a testimony to the vast amount of knowledge about life acquired since the 1953 discovery of the double helix (the same year Miller published the results of his famous experiment). Relevant fields outside of biology also have seen major advances, in supra-molecular chemistry for example, which have greatly expanded our knowledge of the behavior of soft matter in the mesoscopic domain.

Yet despite our extensive knowledge of the molecular biology of the cell, we remain at a complete loss as to what may have been the basic steps which led from the Murchison monomers to the cell system in terms of the known laws of chemistry and physics.

Rather than reveal the elusive path from a chemical soup to the last common ancestor of all extant life, the spectacular progress in cell biology and organic chemistry outlined above has revealed just how immense is the chasm between a soup of organic compounds and the cell with its membrane, the necessary complement of enzyme catalysts, the proteins' synthetic apparatus, genetic information encoded in the double helix, and so forth. Despite many heroic attempts,[19] no one has produced any convincing explanation of how nature could have overcome this chasm, the vastness of which was described recently by Stephen Meyer in his *Signature in the Cell*.[20] (See also Chapter 11 in my own *Evolution: A Theory in Crisis*[21] and Brian Miller's chapter in the newly revised and expanded edition of *The Mystery of Life's Origin*.[22])

The origin-of-life community has identified various steps to the cell. It is widely accepted that four of these are the formation of basic building blocks such as the amino acids and nucleotides; their polymerization into proteins and DNA; the formation of the first primitive replicating system; and the evolution of the modern DNA and protein cell system with a functioning genetic code and an apparatus for protein synthesis. Only work on the first step has seen substantial progress. How the other steps were accomplished in terms of known laws of nature is a complete enigma. The widely acknowledged reality is that within the entire corpus of twenty-first century science, there is no explanation. Science, it seems, has reached an impasse. The origin of life remains as arguably the biggest unsolved problem in science.

In a critical paper summarizing this impasse, specifically with regard to the problem of how the genetic code and translation system could have emerged, Eugene Koonin and Artem Novozhilov offer the following comment:

> At the heart of this problem is a dreary vicious circle: what would be the selective force behind the evolution of the extremely complex translation system before there were functional proteins? And, of course, there could be no proteins without a sufficiently effective translation system. A variety of hypotheses have been proposed in attempts to break the circle but so far none of these seems to be sufficiently coherent or enjoys sufficient support to claim the status of a real theory.[23]

About proto-protein synthesizing systems that were halfway to the modern cell, they comment, "These and other theoretical approaches lack the ability to take the reconstruction of the evolutionary past beyond the complexity threshold that is required to yield functional proteins, and we must admit that concrete ways to cross that horizon are not currently known."[24]

About the transition from an RNA world to the modern DNA/ protein world and its associated models, they write that "we are unaware of any experiments that would have the potential to actually reconstruct the origin of coding, not even at the stage of serious planning."[25]

Summarizing the state of the art they confess to "considerable skepticism." As they explain, "It seems that the two-pronged fundamental question: 'why is the genetic code the way it is and how did it come to be?', that was asked over 50 years ago, at the dawn of molecular biology, might remain pertinent even in another 50 years. Our consolation is that we cannot think of a more fundamental problem in biology."[26]

Overcoming the Impasse

So HOW did the transition from soup to cell occur? One obvious explanation is the idea that an intelligent agency assembled the first cell. This is an explanation popular among some supporters of intelligent design. Although it is rejected by most academic biologists, as the evidence stands it is perhaps as convincing an explanation as any available. However, an alternative possibility (my own preferred position) is that there are new laws, or novel properties of matter yet to be discovered, which enabled the path from chemistry to the cell.

For example, one problem that would have to be overcome in any naturalistic framework is what's known as the clutter problem. In all known prebiotic syntheses, in addition to the desired monomers there is a vast universe of other small, reactive organics. These include a great inventory of closely related amino acids and nucleotides, differing slightly from their biologically important relatives, who are just as likely to join up with the desired monomers, making a polymeric chaos.[27] Getting linear bio-polymers made up of a few basic monomers in a prebiotic setting and avoiding multiple unwanted side reactions with various reactive organics is a real poser. How this could have been achieved prebiotically, before enzymes existed, is a huge problem.

As Gerald Joyce comments, "The chief obstacle to understanding the origin of RNA-based life is identifying a plausible mechanism for overcoming the clutter wrought by prebiotic chemistry."[28] Each of RNA's four components "would have been accompanied by several closely related analogues... which could have assembled in almost any combination."[29] Joyce explains:

> The nucleotides (and their analogues) may even have joined to form polymers, with a combinatorial mixture of 2',5'-, 3',5'- and 5',5'-phosphodiester linkages, a variable number of phosphates between the sugars, D- and L- stereoisomers of the sugars, α- and β-anomers at the glycosidic bond, and assorted modifications of the sugars, phosphates and bases. It is difficult to visualize a mechanism for self-replication that either would be impartial to these compositional differences or would treat them as sequence information in a broader sense and maintain them as heritable features.[30]

In discussing how the clutter problem might be overcome, Joyce continues: "Perhaps there were special conditions that led to the preferential synthesis of activated β-D-nucleotides or the preferential incorporation of these monomers into polymers." He then elaborates:

> For example, the prebiotic synthesis of sugars from formaldehyde can be biased by starting with glycoaldehyde phosphate, leading to ribose 2,4–diphosphate as the predominant pentose sugar... The polymerisation of adenylate, activated as 5'-phosphorimidazolide, yields 2'5'-linked products in solution, but mostly 3'5'-linked products in the presence of a montmorillonite clay. Thus, through a series of biased syntheses, fractionations and other enrichment processes, there may have been a special route to a warm little pond of RNA.[31]

If the origin of life did occur naturally, in some as yet unidentified way, there must be an underlying special fitness in nature to overcome the clutter problem.

Robert Shapiro also anticipated the discovery of new mechanisms and principles. "Self-replicating systems capable of Darwinian evolution appear too complex to have arisen suddenly from a prebiotic soup," he wrote. "This conclusion applies to both nucleic acid systems and to hypothetical protein-based genetic systems. Another evolutionary principle is therefore needed to take us across the gap from mixtures of simple natural chemicals to the first effective replicator. This principle has not been described in detail or demonstrated, but it is anticipated, and given names such as chemical evolution and the self-organization of matter."[32]

Nobel Prize winner Jack Szostak (Medicine, 2009) also expresses the idea that there must be novel phenomena yet to be discovered to explain how life originated. As he and Itay Budin comment, "The discovery of novel physical mechanisms will be essential for a better understanding of how life could have begun."[33]

Paul Davies sounds a similar note in his book *The Fifth Miracle*, though he lays particular emphasis on how radical he suspects the solution will be. "Real progress with the mystery of biogenesis will be made, I believe, not through exotic chemistry, but from something conceptually new,"[34] he comments. He even speculates that perhaps the weird behavior of matter at a subatomic level might have played a role: "Here is a mainstream physical theory that has information at its heart, which it tangles with matter in an intimate way.... Could some sort of quantum-organizing process be just what is needed to explain the origin of informational macromolecules?"[35]

He goes further, connecting all this to the idea of purpose:
Deterministic thinking, even in the weaker forms of de Duve and [Stuart] Kauffman, represents a fundamental challenge to the existing scientific paradigm.... Although biological determinists strongly deny that there is any actual design, or preordained goal, involved in their proposals, the idea that the laws of nature may be slanted towards life, if not contradicting the letter of Darwinism, certainly offends its spirit. It slips an element of teleology back into nature, a century and a half after Darwin banished it.[36]

A paper by Tommaso Bellini and his colleagues seems to hint at something similar when they contrast the Darwinian scenarios, which they term "fantastic luck theories," with fine-tuning (fitness) theories. "Although 'fantastic luck' scenarios are not forbidden by natural laws, they appear increasingly unlikely and hence 'unacceptable' to the sensitivity of the scientists," they write. "As we have seen, the direction taken by the [origin of life] research is to propose scenarios where the 'fantastic luck' is reduced, and replaced by a stronger degree of necessity. How far this could go, how much our existence can instead be viewed as necessary, woven in the deep structure of Nature, is a question that has always interested scientists."[37]

However, at the moment, none of even the most speculative scenarios provides anything beyond the most tentative explanation to avoid the impasse. But this does not mean there will never be a naturalistic explanation. In 1890 no physicist could have conceived of the implications of twentieth-century quantum physics. Which physicist in 1890 could have envisaged wave/particle duality or superposition or the connection of two particles separated by cosmic distances? No one could have imagined the radical implication of the new physics. Given the radical nature of the twentieth-century revolution in physics, the possibility of new discoveries which are currently inconceivable, but which may throw light on the origin-of-life problem and overcome the impasse, cannot be discounted.

A Primal Blueprint

IRRESPECTIVE OF what proximate causes led to the origin of the cell, and irrespective of what may have been the enigmatic steps from the monomers of Murchison to the last common ancestor of all extant life on Earth, the creation of the first carbon-based cell here or, indeed, anywhere in the universe was only possible because of the ensemble of fitness in nature described in this monograph—what I have termed the primal blueprint. As we have seen, this ensemble is in effect a detailed blueprint, integral to the emergence of the carbon-based cell, written into the laws of nature from the beginning, long before it was instantiated in material form.

As we have seen, the blueprint includes:

1. The fitness of the carbon atom to form stable covalent bonds with itself and hydrogen, oxygen, and nitrogen, generating the vast inventory of organic compounds (Chapter 2).

2. The fitness of the directional property of covalent bonds that enable the assembly of large macromolecules with defined 3-D shapes capable of specific biological functions—enzymatic, structural, and genetic (Chapter 3).

3. The fitness of weak bonds to form complementary electrostatic surfaces that can reversibly stick together different parts of macromolecules (such as the two strands of DNA), enzymes to their substrates, and molecular motors to actin fibers (Chapter 3).

4. The fitness of the differences in the electronegativities of carbon, hydrogen, and nitrogen, which leads to the non-polar hydrophobic hydrocarbons and confers on water its polar hydrophilic character. The particular differences in electronegativity among these atoms give us insoluble long-chain hydrocarbons, which form the basis of the cell membrane and enable stable folded proteins (Chapter 4).

5. The fitness of the emergent properties of the cell membrane, including semi-permeability, self-organizing ability, and insulating capacity, properties fit to serve many indispensable biological ends. These include selective adhesion and crawling, and electrical properties that enable the nervous systems of higher organisms (Chapter 4).

6. The unique abilities of phosphates to store and use energy in the aqueous medium of the cell (Chapter 5).

7. The fitness of various metal atoms, including iron and copper, for channeling electrons down electron transport chains and for oxygen handling in hemoglobin and cytochrome c oxidase (Chapters 5 and 6).

8. The fitness of the sodium (Na^+) and potassium (K^+) ions for rapid charge transport across the membrane and the maintenance of the membrane potential (Chapter 6).

9. The fitness of water to serve as the fluid matrix of the cell, including its low viscosity to serve as the medium of circulation, its unsurpassed fitness as a solvent, its hydrophobic force, and its fitness for proton conduction (Chapter 7).

Moreover, the blueprint is not restricted to the generic cell. There are various aspects of the blueprint that appear specifically fit for the cells of higher organisms. For example, aerobic cells of higher organisms depend on the use of biological oxidations to satisfy their energy-hungry metabolism, and this in turn is only possible because of the unique unreactivity of oxygen at ambient temperatures, and the properties of the transition metal atoms such as iron and copper for activating and handling oxygen. The transport of oxygen to the tissues is enabled by the iron atom in hemoglobin. And the transport of CO_2 from the tissues to the lungs depends on the unique properties of the zinc atom in the enzyme carbonic anhydrase.

The cells of complex multicellular organisms also depend on water's low viscosity and the high diffusion rates of solutes in water, which enable the circulatory system as well as metabolically active cells large enough to possess the cytoskeletal machinery to crawl and morph, skills essential in embryogenesis. The high rates of diffusion of solutes in water is also key to the fitness of the small, rapidly diffusing sodium, potassium and chloride ions (Na^+, K^+, and Cl^-) for generating the membrane potential, an indispensable adaptation for transmitting nerve impulses in higher organisms.

Mutual Fitness

WHILE THE properties of the individual atoms are impressively fit for specific biological functions, even more impressive is, as Henderson stressed in *The Fitness of the Environment*, their mutual fitness to work together to achieve various vital ends. In discussing the way water and carbon dioxide work together, Henderson described their association as so close "that it is scarcely correct logically to separate them at all; together they make up the real environment and they never part company."[38]

And again he points out that the fitness of carbon dioxide is "dependent upon water… resting upon solubility and ionization; upon interactions between the two substances."[39] Again he stresses that it is "by many independent and united actions"[40] that the ensembles achieve their life-promoting collective properties.

In these pages we have reviewed various remarkable ensembles of reciprocal fitness, many of them discovered in the decades following Henderson's seminal work. There is the reciprocal fitness of the insoluble non-polar hydrocarbons and the polar hydrophilic nature of water, which gives us the bilayer lipid membrane and the folding of proteins. There is the mutual fitness of the oxygen atom to accept one electron at a time and the ability of transition metals to donate individual electrons, enabling the controlled reduction and activation of oxygen. And there is the mutual fitness of the strong and weak bonds which work together in the formation of the specific shapes and functions of large macromolecules. Those are just three of countless such instances of reciprocal fitness.

Parsimony

FURTHER INTRIGUING evidence of nature's special fitness for the carbon-based cell is the remarkable fact that the same four atoms—hydrogen, carbon, oxygen, and nitrogen—which combine to form the basic organic substances of the cell also provide, by simple combinations with each other, three simple molecules which are vital to life: water (H_2O), the supreme matrix; carbon dioxide (CO_2), the unique carrier of the carbon atom to all parts of the biosphere; and ammonia (NH_3), the compound via which the nitrogen atom is introduced into the organic domain.

It is not just that simple combinations of the four main atoms of organic chemistry provide these three vital molecules but that they are, as we saw above, among the most abundant atoms in the cosmos. Water is formed from the most common atom, hydrogen, and the third most common atom, oxygen. Carbon dioxide (CO_2) is formed from the fourth and third most common atoms in the cosmos, carbon and oxygen. Ammonia (NH_3) is formed from the most common and fifth most common atoms in the universe, hydrogen and nitrogen.

There is yet more. Cells require energy, and where do we find an atom to generate energy by chemical reaction with the substances of life, and with unparalleled efficiency? Again, we need look no further than among these same four atoms. One of them, oxygen, the third most common atom in the cosmos, is uniquely fit to provide copious quantities of energy for living systems. No other atom comes close. Additionally, the oxidation of carbon and hydrogen releases greater amounts of energy than any other type of oxidation.[41] So two of the same four atoms—carbon and hydrogen—serve as maximally fit stores of metabolic energy.

Has any artifact of man ever transcended the wondrous elegance and parsimony of the ensemble of fitness embedded in these four atoms? Only someone committed to rejecting out of hand all evidence of teleology in nature could fail to see in these ensembles of fitness and in this elegant parsimony evidence of design.

Alien Life

THE PRIMAL blueprint means that nature is uniquely fit for the cell as it exists on Earth, to the almost certain exclusion of any other hypothetical forms of chemical life, natural or artificial. And this is predictive in an important sense: it suggests that all chemical life throughout the cosmos will be carbon-based and resemble life on Earth; or at the very least it suggests that any life found elsewhere in the cosmos that even approaches the sophistication of life on Earth will be based on the same ensemble of carbon and its primary chemical collaborators.

As Henderson noted more than a century ago, "Carbon, hydrogen, and oxygen, each by itself, and all taken together, possess unique and preeminent chemical fitness for the organic mechanism."[42] The fitness of this ensemble, he adds, "results from characteristics which constitute a series of maxima—unique or nearly unique properties of water, carbonic acid, the compounds of carbon, hydrogen, and oxygen...—so numerous, so varied, so nearly complete among all things which are concerned in the problem that together they form certainly the greatest possible fitness... to promote complexity, durability, and active metabolism in the organic mechanism which we call life."[43]

He asks, "What are the possibilities of obtaining the same characteristics in other substances?"[44] His answer, "No other environment... could possess a like number of fit characteristics" for promoting "the organic mechanism we call life."[45]

Anyone familiar with the current state of astrobiology knows that no other set of atoms has ever been discovered or even imagined that could equal the mutual fitness of the members of the ensemble that underlies the existence of carbon-based life.

Astrobiologists are guided by NASA's motto, "follow the water," in their search for life beyond Earth. And this amounts to a tacit admission of the lack of a suitable alternative solvent. Astrobiologists Frances Westall and André Brack call liquid water the "perfect solute for carbon-based life." And they don't stop at water. They add: "Schulze-Makuch and Irwin (2004) make an excellent discussion about the possibility of life based on other chemical compounds and solutes but none of these combinations presents the same range of properties and advantages as carbon and water and, at least on Earth, most are not stable or are outcompeted by carbon."[46]

In the journal *Astrobiology*, Christopher McKay lists the following as minimum requirements for a life-detection mission: "(1) Liquid water of suitable salinity, past or present; (2) Carbon in the water; (3) Biologically available N in the water; (4) Biologically useful energy in the water; (5) Organic material that can possibly be of biological origin and a plausible strategy for sampling this material."[47]

One century after Henderson, no fact has come to light to seriously threaten his conclusion that the laws of nature are finely tuned for life as it exists on Earth. On the contrary, many discoveries and advances unimagined in 1913 have further confirmed it. These include the importance of the hydrophobic force, the utility of strong covalent and weak bonds for the assembly of complex macromolecules, proton conduction, and the fitness of ten or so metal atoms for very specific cellular functions. Further confirming Henderson's conclusion are advances outside the biological sciences, such as the elucidation of the cosmic abundances of the atoms, the carbon-12 resonance that enables stars to generate carbon, and the fine-tuning of the constants of physics for life.

Darwinism Diminished

ANOTHER IMPLICATION of the primal blueprint is the obvious challenge it poses to the Darwinian world view. The existence of the blueprint specifying the design of the cell long before life emerged on Earth implies that, however the first cell was assembled, whether suddenly through the direct agency of a Divine watchmaker, or gradually by a blind "Darwinian" watchmaker (constrained by natural law and utilizing the natural selection of chance events), or by some other as yet unimagined means, its creation was only possible because the blueprint was already in place.

The fitness of the atoms and the makeup of the primal blueprint for the cell was determined long before the Earth was born. Consequently, even if we posit a blind Darwinian watchmaker as the major player in the actualization of the blueprint, its actualization was only possible because the blueprint was already in place.

In Sum

WE HAVE seen in this monograph that the miracle of the cell is only possible because of a vast prior ensemble of mutual fitness in the unique properties of a set of about one-fifth of the atoms of the periodic table. There is, I believe, no other corpus of evidence anywhere in science which provides more convincing evidence of design in nature and purpose in the universe than the fine tuning of these atoms for the carbon-based cell. On any consideration, the evidence presented in this monograph conveys the irresistible impression that the properties of the atoms have been contrived directly and purposefully to enable the existence of life in the universe.

As far as the actualization of the blueprint is concerned, which remains the deepest of mysteries, I believe that further elements of fitness in nature will be discovered over the coming decades which will finally reveal the fateful path from chemistry to life. And I believe that when the path is finally elucidated, it will turn out to be extraordinary, one of the greatest of scientific wonders, revealing a far deeper teleology in nature than all the elements of natural fitness for the cell and life documented so far.

Even more, I believe that the elucidation of that fateful route will be of far greater intellectual consequence than any other discovery in science since the birth of science in the sixteenth century. Indeed, I believe the path, when discovered, will prove to be so obviously indicative of a profound teleology in the very ground of being that it will prove a watershed in the history of thought.

Conversely, if instead it is eventually established that there is no purely natural path across the great gulf from non-life to life, and that only the additional exertion of an intelligent agent could have assembled the first cell on Earth, that will be equally a watershed in human thought.

Finally, even if the design inference is rejected, the existence of the primal blueprint for the cell, revealed as a result of scientific discoveries over the past two centuries, provides what is surely irrefutable evidence that the cosmos as currently understood is uniquely fit for carbon-based life, and that life as it exists on Earth and beings of our biological design occupy a very special place in nature. Irrespective of any design inference, what science has revealed already confirms the deep intuition of the medieval Christian scholars who believed that "in the cognition of nature in all her depths, man finds himself."[48]

Endnotes

Introduction

1. Carl Sagan, on *Cosmos: A Personal Voyage*, episode 12, "Encyclopaedia Galactica," directed by Adrian Malone, written by Carl Sagan, Ann Druyan, and Steven Soter, aired December 14, 1980, on PBS.

1. The Amazing Cell

1. David Rogers, "Neutrophil Chasing Bacteria," Embryology Education and Research, video, 0:33, accessed May 18, 2020, https://embryology.med.unsw.edu.au/embryology/index.php/Movie_-_Neutrophil_chasing_bacteria.

2. Goethe's Faustus exclaimed, "Mysterious even in open day, Nature retains her veil, despite our clamors: That which she doth not willingly display, Cannot be wrenched from her with levers, screws, and hammers." Johann Wolfgang von Goethe, *Faust*, in *Goethe's Faust, With Some of the Minor Poems*, ed. and trans. Elizabeth Craigmyle (London: W. J. Gage, 1889), 31.

3. Erika Check Hayden, "Human Genome at Ten: Life is Complicated," *Nature* 464 (2010): 664–667.

4. Howard C. Berg, "Bacterial Microprocessing," *Cold Spring Harbor Symposium on Quantitative Biology* 55 (1990): 539.

5. Rob Phillips, *Physical Biology of the Cell*, 2nd ed. (New York: Garland Science, 2013), chap. 3.

6. Vance Tartar, "Regeneration," in *The Biology of Stentor* (London: Pergamon Press, 1961), 105–135.

7. Herbert Spencer Jennings, *Behavior of the Lower Organisms* (New York: The Columbia University Press, 1906), 336–337.

8. Kevin B. Clark, "Origins of Learned Reciprocity in Solitary Ciliates Searching Groups 'Courting' Assurances at Quantum Efficiencies," *Biosystems* 99 (2010): 27–41.

9. Jennings, *Behavior of the Lower Organisms*, 337. And he continues, "We usually do attribute consciousness to the dog, because this is useful; it enables us practically to appreciate, foresee, and control its actions much more readily than we could otherwise do so. If Amoeba were so large as to come within our everyday ken, I believe it beyond question that we should find similar attribution to it of certain states of consciousness a practical assistance in foreseeing and controlling its behaviour… it may perhaps be said that objective investigation is as favorable to the view of the general distribution of consciousness throughout the animals as it could well be."

10. Brian J. Ford, "The Secret Power of the Single Cell," *New Scientist* 2757 (2010): 26.

11. Ford, "The Secret Power," 26. Sentience in higher organisms has long urged the question, just how far down the evolutionary tree might sentience reach? See Hans Jonas, *The Phenomenon of Life* (Evanston, IL: Northwestern University Press, 1966). Jonas argues on grounds of evolutionary continuity that there is no compelling reason to stop until we reach the simplest life: "The *continuity* of descent now established between man and the animal world made it impossible any longer to regard his mind, and mental phenomena as such, as the abrupt ingression of an ontologically foreign principle at just this point [in the evolutionary process]... Where else than at the beginning of life can the beginning of inwardness be placed?" (57–58). The neurologist Charles Sherrington in *Man on His Nature* (Cambridge: Cambridge University Press, 1963), also saw that mind as we experience it could hardly be present in a single-celled organism, but added, "Not that there would seem any inherent unlikelihood in mind attaching in some degree to an individual consisting of a single cell" (148). True, single-celled organisms lack any recognizable nervous system, but Sherrington insists that "the cell framework, the cytoskeleton, might serve" and that there is "no need for our imagination to halt and say 'the apparatus for it is wanting.'" He concludes, "There seems no lower limit to mind" (265). The primatologist Robert Yerkes also suggested that microorganisms might possess rudimentary sentience; see Robert M. Yerkes, "Animal Psychology and Criteria of the Psychic," *Journal of Philosophy, Psychology and Scientific Methods* 2, no. 6 (1905): 141–149.

12. Jacques Monod, *Chance and Necessity* (London: Collins, 1972), 64.

13. Lawrence Henderson, *The Fitness of the Environment* (New York: Macmillan, 1913), 312.

2. The Chosen Atom

1. Arthur E. Needham, prefatory poem to *The Uniqueness of Biological Materials* (London: Pergamon Press, 1965), v.

2. David Emmite, Moshe Sipper, and James A. Reggia, "Go Forth and Replicate," *Scientific American* 285 (2001): 35.

3. Lawrence Henderson, *The Fitness of the Environment* (New York: MacMillan, 1913), 191, https://archive.org/details/cu31924003093659/page/n209.

4. William Prout, *Chemistry, Meteorology, and the Function of Digestion Considered with Reference to Natural Theology*, The Bridgewater Treatises 8 (London: William Pickering, 1834), 436.

5. Prout, *Chemistry*, 436.

6. As Peter Mark Roget said, "Thus a degree of heat, which would occasion no change in most mineral [inorganic] substances, will at once effect the complete disunion of the elements of an animal or vegetable body. Organic substances are, in like manner, unable to resist the slower, but equally destructive agency of water and atmospheric air [oxygen]; and they are also liable to various spontaneous changes, such as those constituting fermentation and putrefaction, which occur when their vitality is extinct, and when they are consequently abandoned to the uncontrolled operation of their natural chemical affinities. This tendency to decomposition may, indeed, be regarded as inherent in all organized substances, and as requiring for its counteraction, in the living system, that perpetual renovation of materials which is supplied by the powers of nutrition" (8). "The materials which nature has employed in [the construction of organic substances]... are of a more plastic quality [compared with mineral or inorganic substances], and which allow of a variable proportion of ingredients, and of great diversity in the modes of their combination" (4–5). Peter Mark Roget, *Animal and Vegetable Physiology Considered with Reference to Natural Theology*, Bridge-

water Treatises 5, vol. 2 (London: William Pickering, 1834), https://archive.org/details/pt2bridgewatertr05londuoft. Peter Ramberg commented recently, "Organic compounds, derived from plant and animal sources, were less stable, more prone to decomposition, and had compositions more difficult to ascertain by elemental analysis." Peter J. Ramberg, "The Death of Vitalism and the Birth of Organic Chemistry: Wöhler's Urea Synthesis and the Disciplinary Identity of Organic Chemistry," *Ambix* 47, no. 3 (November 2000): 170.

7. William Prout was impressed by the variety and diversity of organic compounds. He spoke of the "countless forms and varieties" (Prout, *Chemistry*, 440) of organic compounds containing carbon. And in another passage he comments, "Carbon perhaps more than any other principle may be considered as constituting the staminal or fundamental element entering into the composition of organized beings. This is particularly the case in principles [carbon compounds] from the vegetable kingdom, which owe their peculiar character essentially to carbon, and their endless varieties to differences in its quantity, and to the modifying influence of the hydrogen and oxygen with which it is associated. In animal substances carbon exerts a similar influence" (103).

8. Roget, *Animal and Vegetable*, "So great is their complexity, that... no human art is adequate to effect their reunion" (5). Other authors were similarly impressed by their complexity compared with inorganic compounds. J. L. Comstock spoke of them being "distinguished from the mineral creation... by their more complicated nature." See John Lee Comstock, *Conversations on Chemistry: In Which the Elements of That Science Are Familiarly Explained, and Illustrated by Experiments* (Hartford, CT: Beach and Beckwith, 1835), 239.

9. Isaac Asimov, *The World of Carbon* (New York: Collier Books, 1968), 11–12. Francis Preston Venable explained an empirical basis for early nineteenth-century vitalism in his *Short History of Chemistry* (Boston: D. C. Heath, 1907), 118. He pointed out that vitalists of the early nineteenth century thought that while "mineral substances could be artificially produced, or synthesized, the imitation of organic bodies was beyond the reach of experimental methods, as they were the products of life itself, and could be formed only in the plant or animal cell. It is true that new organic preparations had been made by distilling and otherwise treating various products of plant life, but the original source or starting-point remained the same life products. [However,] Chevreul had shown that the natural fats were compounds of certain acids and the glycerine discovered by Scheele. Still, all of this did not do away with the belief in the necessity for the action of the mysterious life force." See also Ramberg, "The Death of Vitalism," where he comments: "Whereas inorganic compounds... were easily analyzed and synthesized, organic compounds could be made only in plants or animals by a mysterious vital force that could not be replicated in the laboratory" (170).

10. Prout wrote, "The attainment of the particular purpose of organic life is effected, not by any departure from the great scheme, but by new and different combinations. To suppose... that... oxygen and hydrogen [can be combined in an organism], in exactly the same proportion, and in the same manner, in which they are combined, when they exist as water; and, from these elements so combined, can yet produce something different from water, is contrary to all reason... organized substances are composed of the same elements, which exist abundantly throughout the world in the unorganized state; moreover... these elements are subject to all the influences and agencies of inorganic nature" (Prout, *Chemistry*, 433–434). He also commented, "We cannot, however, produce artificially either sugar, or any other organic compound, by directly combining their elements; because we cannot bring their elements together, precisely in the requisite state and proportions. Still, there is no doubt,

that if the elements could be so brought together, the compound thence resulting, would be the same as the natural compound" (418).

11. Friedrich Wöhler to Jöns Jacob Berzelius, February 22, 1828, in *Briefwechsel zwischen J. Berzelius und F. Wöhler*, vol. 1, ed. O. Wallach, trans. W. H. Brock (Leipzig: Verlag von Wilhelm Engelmann, 1901), 206, cited at "Friedrich Wöhler," Today in Science History, accessed May 29, 2019, https://todayinsci.com/W/Wohler_Friedrich/WohlerFriedrich-Quotations.htm.

12. Venable, *Short History of Chemistry*, 118.

13. Alan J. Rocke, "Hermann Kolbe," *Encyclopaedia Britannica*, accessed May 22, 2019, https://www.britannica.com/biography/Hermann-Kolbe.

14. Henderson, *Fitness*, 192.

15. See, for example, Thomas Nagel, *Mind and Cosmos: Why the Materialist Neo-Darwinian Conception of Nature Is Almost Certainly False* (New York: Oxford University Press, 2012); and J. Scott Turner, *Purpose and Desire: What Makes Something "Alive" and Why Modern Darwinism Has Failed to Explain It* (New York: HarperOne, 2017).

16. Henderson, *Fitness*, 193.

17. J. J. C. Mulder, "Theoretical Organic Chemistry: Looking Back in Wonder," in *Theoretical Organic Chemistry*, ed. C. Párkányi (New York: Elsevier, 1997), 1–32.

18. Isaac Asimov, *A Short History of Chemistry* (New York: Anchor Books, 1965), 98–99.

19. Henderson, *Fitness*, 193.

20. Alfred Russel Wallace, *Man's Place in the Universe* (London: Chapman and Hall, 1904), https://archive.org/details/manuniverse00walluoft/page/n209.

21. N. V. Sidgwick, *The Chemical Elements and Their Compounds*, vol. 1 (Oxford: Oxford University Press, 1950), 490. "In the first place, the typical 4-covalent state of the carbon atom is one in which all the formal elements of stability are combined. It has an octet, a fully shared octet, an inert gas number, and in addition, unlike all the other elements of the group, an octet which cannot increase beyond 8, since 4 is the maximum covalency possible for carbon. Hence the saturated carbon atom cannot co-ordinate either as donor or as acceptor, and since by far the commonest method of reaction is through co-ordination, carbon is necessarily very slow to react, and even in a thermodynamically unstable molecule may actually persist for a long time unchanged. More than 50 years ago Victor Meyer drew attention to the characteristic 'inertness' (*Trägheit*) of carbon in its compounds, and there can be no doubt that this is its main cause."

22. P. W. Atkins, *The Periodic Kingdom: A Journey into the Land of the Chemical Elements* (New York: Basic Books, 1995), 17.

23. Atkins, *The Periodic Kingdom*, 17.

24. Atkins, *The Periodic Kingdom*, 17.

25. Asimov, *The World of Carbon*, 14.

26. Primo Levi, *The Periodic Table* (London: Abacus, 1990), 226–227. It's true that silicon can bond with itself into long chains, but the bonds are less stable, particularly in the presence of not just oxygen but also water as well as ammonia, sometimes proposed as an alternative to water. In Needham's *The Uniqueness of Biological Materials* (page 37), he gives the strength of C-C bonds as twice that of S-S bonds. And see my discussion of silicon life in my article "The Place of Life and Man in Nature: Defending the Anthropocentric Thesis," *BIO-Complexity* 2013 (1): 1–15, https://doi.org/10.5048/bio-c.2013.1.

27. Needham, *Uniqueness*, 30.

28. George Wald, "The Origins of Life," *PNAS* 52 (1964): 594–611.

29. For a simple, lucid account of the carbon double bond and why silicon can't form such bonds, see S. E. Gould, "Shine On You Crazy Diamond: Why Humans are Carbon-Based Lifeforms," *Scientific American*, November 11, 2012, https://blogs.scientificamerican.com/lab-rat/shine-on-you-crazy-diamond-why-humans-are-carbon-based-lifeforms/. Also see Kira Mitsuo, "Bonding and Structure of Disilenes and Related Unsaturated Group-14 Element Compounds," *Proceedings of the Japan Academy, Series B* 88, no. 5 (2012): 167–191, https://doi.org/10.2183/pjab.88.167.

30. Bruce Alberts et al., "Catalysis and the Use of Energy by Cells," in *Molecular Biology of the Cell*, 4th ed. (New York: Garland Science, 2002), https://www.ncbi.nlm.nih.gov/books/NBK26838/; and see Alberts et al., "Protein Function," in *Molecular Biology of the Cell*, https://www.ncbi.nlm.nih.gov/books/NBK26911/: "Extremely high rates of chemical reaction are achieved by enzymes—far higher than for any synthetic catalysts. This efficiency is attributable to several factors. The enzyme serves, first, to increase the local concentration of substrate molecules at the catalytic site and to hold all the appropriate atoms in the correct orientation for the reaction that is to follow. More importantly, however, some of the binding energy contributes directly to the catalysis. Substrate molecules must pass through a series of intermediate states of altered geometry and electron distribution before they form the ultimate products of the reaction. The free energy required to attain the most unstable transition state is called the activation energy for the reaction, and it is the major determinant of the reaction rate. Enzymes have a much higher affinity for the transition state of the substrate than they have for the stable form. Because this tight binding greatly lowers the energies of the transition state, the enzyme greatly accelerates a particular reaction by lowering the activation energy that is required" [internal references removed].

31. Alberts et al., "Catalysis and the Use of Energy by Cells."

32. Many sources give the energies involved in conformational changes in proteins. These are about 10–20 kcal/mol. K. N. Houk et al., "Binding Affinities of Host–Guest, Protein–Ligand, and Protein–Transition-State Complexes," *Angewandte Chemie International Edition* 42, no. 40 (October 20, 2003): 4872–4897, https://doi.org/10.1002/anie.200200565. Panagiotis L. Kastritis et al., "A Structure-Based Benchmark for Protein-Protein Binding Affinity: Protein-Protein Structure-Affinity Benchmark," *Protein Science* 20, no. 3 (March 2011): 482–491, https://doi.org/10.1002/pro.580.

33. Bond energies in kJ/mol are given as Na-Cl 787, Na–F, 923, in "Lattice Energy," College of Science: Chemical Education Division Group, Purdue University, accessed May 8, 2019, http://chemed.chem.purdue.edu/genchem/topicreview/bp/ch7/lattice.html; C-H 413 is given in Kim Song and Donald Le, "Bond Energies," LibreTexts Chemistry, accessed May 30, 2019, https://chem.libretexts.org/Bookshelves/Physical_and_Theoretical_Chemistry_Textbook_Maps/Supplemental_Modules_(Physical_and_Theoretical_Chemistry)/Chemical_Bonding/Fundamentals_of_Chemical_Bonding/Bond_Energies. The greater stability of inorganic materials than organic materials can be seen in the fact that they only undergo chemical reaction (involving the breaking of their chemical bonds) at far higher temperatures, which cause much more energetic collisions between molecules, imparting (on collision) sufficient energy to overcome activation energy barriers.

34. John Gribbin, *Cosmic Coincidences: Dark Matter, Mankind and Anthropic Cosmology* (London: Black Swan, 1991), 14.

35. Needham, *Uniqueness*, 30.

36. Needham, *Uniqueness*, 30.

37. Henderson, *Fitness*, 220.

38. Sidgwick, *The Chemical Elements*, 490.

39. Robert E. D. Clark, *The Universe: Plan or Accident?*, 3rd ed. (Grand Rapids: Zondervan, 1972), 97.

40. K. W. Plaxco and M. Gross, *Astrobiology: A Brief Introduction*, 2nd ed. (Baltimore: Johns Hopkins University Press, 2011), 12.

41. Lynn J. Rothschild and Rocco L. Mancinelli, "Life in Extreme Environments," *Nature* 409, no. 6823 (February 22, 2001): 1092–1101, https://doi.org/10.1038/35059215.

42. Stanley L. Miller and Leslie E. Orgel, *The Origins of Life on the Earth* (Upper Saddle River, NJ: Prentice Hall, 1974). See chap. 9 on the stability of organic compounds.

43. Miller and Orgel, *The Origins of Life on the Earth*.

44. Michael Denton, *Nature's Destiny: How the Laws of Biology Reveal Purpose in the Universe* (New York: Free Press, 1998), 110. See also T. Hoyem and O. Kvale, *Physical, Chemical and Biological Changes in Food Caused by Thermal Processing* (London: Applied Science Publishers, 1977), 185–201.

45. H. R. White, "Hydrolytic Stability of Biomolecules at High Temperatures and Its Implication for Life at 250°C," *Nature* 310 (1984): 430–432. See also H. Bernhardt, D. Ludeman, and R. Jaenicke, "Biomolecules Are Unstable under Black Smoker Conditions," *Naturwissenchaften* 71 (1984): 583–586.

46. White, "Hydrolytic Stability of Biomolecules," 430–432. See also Bernhardt, Ludeman, and Jaenicke, "Biomolecules," 583–586.

47. Andrew Clarke et al., "A Low Temperature Limit for Life on Earth," *PloS One* 8, no. 6 (2013): e66207, https://doi.org/10.1371/journal.pone.0066207.

48. Figure data from Leon M. Lederman, *From Quarks to Cosmos* (New York: Scientific American Library, 1989), 152. And see F. Hoyle, "Ultra High Temperatures," *Scientific American* 191, no. 3 (1954): 145–154.

49. Water is liquid between 0C° and 100C° at an atmospheric pressure of 760 mm Hg. At higher atmospheric pressures water can exist as a liquid at temperatures considerably above 100C°.

50. Henderson, *Fitness*. See also Sidgwick, *The Chemical Elements*, 490; and Needham, *Uniqueness*. See Clark, *The Universe: Plan or Accident?*, chap. 8: "No other element can compete... No other element is like it.... Carbon is in fact unique." See also Wald, "The Origins of Life"; and C. F. A. Pantin, *The Relations Between the Sciences* (Cambridge, UK: Cambridge University Press, 1968), Appendix 1, 145: "The organic chemistry of carbon is uniquely suitable for the construction of complex systems... no other element provides such a necessary and unique collection of properties.... if life has arisen elsewhere in the universe... it would be likely to be a water–carbon compound system." Also see George Wald, "Fitness in the Universe: Choices and Necessities," *Origins Life* 5 (1974): 7–27; and Harold J. Morowitz, *Cosmic Joy and Local Pain: Musings of a Mystic Scientist* (New York: Scribner, 1987). Harold Morowitz, at the Harvard-Smithsonian Center for Astrophysics, Boston, confirmed in a personal communication in 2003 his belief in the uniqueness

of the carbon atom for chemical life. See also John D. Barrow, *The Anthropic Cosmological Principle* (Oxford: Oxford University Press, 1988); and David C. Catling et al., "Why O_2 Is Required by Complex Life on Habitable Planets and the Concept of Planetary 'Oxygenation Time,'" *Astrobiology* 5, no. 3 (June 2005): 415–438. Lynn Rothschild writes, "While silicon is also common (though not nearly as common as carbon in the Universe as a whole) and can form interesting polymers, its flexibility pales in comparison with organic chemistry, particularly in the ability of carbon to form polymers. Most stunning of all, organic compounds—including amino acids and nucleotide bases—have been detected in the interstellar medium. Even Earth, which is composed of a substantial quantity of silicates and thus should be biased towards silicon-based life, harbours carbon-based life" [internal references removed]. See Lynn J. Rothschild, "The Evolution of Photosynthesis... Again?," *Philosophical Transactions of the Royal Society B: Biological Sciences* 363, no. 1504 (August 27, 2008): 2787–2801. See also N. R. Pace, "The Universal Nature of Biochemistry," *PNAS* 98 (2001): 805–808; J•Gale, *Astrobiology of Earth: The Emergence, Evolution, and Future of Life on a Planet in Turmoil* (Oxford: Oxford University Press, 2009); L. N. Irwin and D. Schulze-Makuch, *Cosmic Biology: How Life Could Evolve on Other Worlds* (New York: Praxis and Springer, 2011); Kevin W. Plaxco, *Astrobiology: A Brief Introduction*, 2nd ed. (Baltimore: Johns Hopkins University Press, 2011), chap. 1; David L. Abel, "Is Life Unique?," *Life* 2, no. 1 (2012): 106–134; and D. Schulze-Makuch and L. N. Irwin, "The Prospect of Alien Life in Exotic Forms on Other Worlds," *Naturwissenschaften* 93 (2006): 155–172. Even Carl Sagan conceded he was a carbon-and-water chauvinist; see Carl Sagan, *Cosmos* (New York: Ballantine Books, 1985), 105.

51. Plaxco and Gross, *Astrobiology*, 6.

52. Alfred Russel Wallace, *The World of Life: A Manifestation of Creative Power, Directive Mind and Ultimate Purpose* (London: Chapman and Hall, 1911), 393.

3. THE DOUBLE HELIX

1. Horace Judson, *The Eighth Day of Creation: Makers of the Revolution in Biology* (New York: Simon and Schuster, 1979), 173–175.

2. James D. Watson, *Molecular Biology of the Gene*, 3rd ed. (California: W. A. Benjamin, 1976), 25–28.

3. Watson, *Molecular Biology*, 28.

4. Watson, *Molecular Biology*, 28.

5. Watson, *Molecular Biology*, 28.

6. Watson, *Molecular Biology*, 28.

7. John Cairns, Gunther S. Stent, and James D. Watson, *Phage and the Origins of Molecular Biology* (New York: Cold Spring Harbor Laboratory Press, 2007). Other important researchers associated with the phage group include Salvador Luria, James Watson, Alfred Hershey, Gunther Stent, Frank Stahl, Seymour Benzer, and Renato Dulbecco.

8. Judson, *The Eighth Day of Creation*, 60.

9. Proteins have long been invested with a mystical aura. Even the word "protein," proposed in 1838 by the Swedish chemist Berzelius in a letter to his student Gerrit Jan Mulder, derives from the Greek meaning "holding first place," conveying the idea that these molecules play a unique and vital role in biology. See Joseph S. Fruton, *Proteins, Enzymes, Genes* (New Haven, CT: Yale University Press, 1999), 171. For more information, see Graeme K. Hunter, *Vital Forces* (New York: Academic Press, 2000), chap. 11.

10. Maclyn McCarty, "Discovering Genes Are Made of DNA," *Nature* 421 (2003): 406.

11. M. Delbrück, "A Physicist's Renewed Look at Biology: Twenty Years Later," *Science* 168, no. 3937 (1970): 1312.

12. Francis Crick, "The Impact of Linus Pauling on Molecular Biology" (lecture, Pauling symposium, Oregon State University, 1995). A transcript of Crick's talk is available at "The Life and Work of Linus Pauling (1901–1994: A Discourse on the Art of Biogeography," Special Collections & Archives Research Center, Oregon State University Libraries, 1995, http://scarc.library.oregonstate.edu/events/1995paulingconference/video-s1-2-crick.html. See also Judson, *The Eighth Day of Creation*, section 1, chap. 1, part c.

13. Erwin Schrödinger, *What Is Life? The Physical Aspect of the Living Cell, with Mind and Matter & Autobiographical Sketches* [1944] (Cambridge: Cambridge University Press, 1992), 68.

14. Jacques Monod, *Chance and Necessity* (New York: Vintage Books,1972), 61. It is by virtue of their capacity to form, with other molecules, stereospecific and non-covalent complexes that proteins exercise their "demoniacal" functions.

15. J. C. Kendrew et al. [1958], quoted in M. F. Perutz, "X-Ray Analysis, Structure and Function of Enzymes," *European Journal of Biochemistry* 8 (1969): 455, https://doi.org/10.1111/j.1432-1033.1969.tb00549.x.

16. Judson, foreword to *The Eighth Day of Creation*, 1st ed., 12.

17. Heinz Neumann et al., "Encoding Multiple Unnatural Amino Acids via Evolution of a Quadruplet-Decoding Ribosome," *Nature* 464, no. 7287 (March 18, 2010): 441–444.

18. Vitor B. Pinheiro et al., "Synthetic Genetic Polymers Capable of Heredity and Evolution," *Science* 336, no. 6079 (April 20, 2012): 341–344.

19. In the *Fitness of the Environment*, Lawrence Henderson defined organisms as durable (self-maintaining and self-replicating) physiochemical mechanisms that exhibit a high degree of complexity and are capable of regulation and homeostasis via metabolic processes which involve an exchange of matter and energy with their surroundings. See Henderson, *The Fitness of the Environment* (New York: MacMillan, 1913), 32–35. Available at https://archive.org/details/cu31924003093659/page/n209. Since Henderson's day, many other authors have formulated definitions of life and organisms (see David L. Abel, "Is Life Unique?," *Life* 2, no. 1 (2012): 106–134), but most include the basic elements identified by Henderson. See John A. Baross et al., *The Limits of Organic Life in Planetary Systems* (Washington, DC: National Academies Press, 2007); Joseph Gale, *Astrobiology of Earth: The Emergence, Evolution, and Future of Life on a Planet in Turmoil* (Oxford: Oxford University Press, 2009), chap. 1; Norman R. Pace, "The Universal Nature of Biochemistry," *PNAS* 98 (2001); Kevin W. Plaxco and Michael Gross, *Astrobiology: A Brief Introduction*, 2nd ed. (Baltimore: Johns Hopkins University Press, 2011), chap. 1; D. Schulze-Makuch and L. N. Irwin, "The Prospect of Alien Life in Exotic Form on Other Worlds," *Naturwissenschaften* 93 (2006): 155–172. One additional element mentioned by recent researchers has been the requirement for a bounding membrane to separate the organism from its environment. For example, see Gerald Joyce, who has defined life as that which can undergo Darwinian evolution (implying that replication must be error-prone to some degree), and Abel, "Is Life Unique?," 107. This requirement is also stressed in Steven A. Benner, Alonso Ricardo, and Matthew A. Carrigan, "Is There a Common Chemical Model for Life in the Universe?," *Current Opinions in Chemical Biology* 8, no. 6 (2004): 672–689, and adopted by NASA (Baross et al., *The Limits of Organic Life*, 1–2). Astrobiologists Irwin and Schultze-Makuch describe life as consisting "of a highly organized system separated by physical boundaries

from its more disordered surroundings" which "maintains its high level of organization and performs work by transforming energy from its environment in a self-regulating manner" and that reproduces "itself autonomously by assembling raw materials from its environment into near-exact replicas according to information perpetuated through an indefinite succession of generations," in D. Schulze-Makuch and L. N. Irwin, *Cosmic Biology* (New York: Springer-Praxis, 2011), 42.

20. Daniel E. Koshland, "The Mechanism of Enzymatic Action," *Encyclopaedia Britannica*, accessed May 9, 2019, https://www.britannica.com/science/protein/The-mechanism-of-enzymatic-action (page 19 of e-version): "Enzymes are more efficient than human-made catalysts operating under the same conditions. Because many enzymes with different specificities occur in a cell, adequate space exists only for a few enzyme molecules catalyzing one specific reaction. Each enzyme, therefore, must be very efficient." A bit later he adds, "The reason for the great efficiency of enzymes is not completely understood. It results in part from the precise positioning of the substrates and the catalytic groups at the active site, which serves to increase the probability of collision between the reacting atoms. In addition, the environment at the active site may be favourable for reaction—that is, acidic and basic groups may act together more effectively there, or some strain may be induced in the substrate molecules so that their bonds are broken more easily, or the orientation of the reacting substrates may be optimal at the enzyme surface. The theories that have been formulated to account for the high catalytic efficiency of enzymes, although reasonable, still remain to be proved."

21. P. W. Atkins, *The Periodic Kingdom: A Journey into the Land of the Chemical Elements* (New York: Basic Books, 1995), 177–178.

22. See Peter Atkins and Loretta Jones, *Chemistry: Molecules, Matter and Change* (New York: W. H. Freeman, 1997), 294–295.

23. Robert E. D. Clark, *The Universe: Plan or Accident?*, 3rd ed. (Grand Rapids: Zondervan Publishing House, 1972), 94. [emphasis in original]

24. Atkins, *The Periodic Kingdom*, 178.

25. Watson, *Molecular Biology*, 90.

26. Watson, *Molecular Biology*, 86.

27. Crick, "The Impact of Linus Pauling."

28. Watson, *Molecular Biology*, 90.

29. Watson, *Molecular Biology*, 98.

30. Watson, *Molecular Biology*, 97–98.

31. Specific binding of a particular substrate (key) to its binding site on the protein (lock) depends on several bonds being arranged in very precise complementary spatial positions in both the substrate and the binding site. Decreasing the number of weak bonds to compensate for their hypothetical increased strength would not work, since specificity of binding would be lost.

32. Watson, *Molecular Biology*, 100.

33. Rob Phillips, *Physical Biology of the Cell*, 2nd ed. (New York: Garland Science, 2013), 127. For more on non-covalent bonds, see Harvey Lodish et al., "Section 2.2: Noncovalent Bonds," in *Molecular Biology*, 4th ed. (New York: W. H. Freeman, 2000), https://www.ncbi.nlm.nih.gov/books/NBK21726/.

34. Energy of ultra-low-frequency (3,000 Hertz) radio photon (calculated by E = Planck constant × Hertz), is $6.63 \times 10^{-34} \times 3,000 = 2 \times 10^{-30}$ J. Energy of 4 kJ/mol hydrogen bond, converted from kJ/mol to joules, is 4×10^3 J / 6.022×10^{23} molecules/mol = 6.7×10^{-21} J. Energy of Big Bang given by NASA at https://web.archive.org/web/20140819120709/ http://imagine.gsfc.nasa.gov/docs/ask_astro/answers/980211b.html.

35. James D. Watson, preface to *DNA: The Secret of Life* (London: William Heinemann, 2003), xii–xiii.

36. Daniel Dennett, *Sweet Dreams: Philosophical Obstacles to a Science of Consciousness* (Cambridge, MA: MIT Press, 2005), 178.

4. CARBON'S COLLABORATORS

1. Peter W. Atkins, *The Periodic Kingdom: A Journey into the Land of the Chemical Elements* (New York: Basic Books, 1995), 149.

2. George Wald, "The Origins of Life," *PNAS* 52 (1964): 595–610.

3. See Lawrence Henderson, *The Fitness of the Environment* (New York: MacMillan, 1913), 210–211. Available online here: https://archive.org/details/cu31924003093659/page/ n209.

4. Henderson, *Fitness*, 219.

5. Atkins, *The Periodic Kingdom*, 20.

6. Atkins, *The Periodic Kingdom*, 20–21.

7. Atkins, *The Periodic Kingdom*, 21.

8. As has been discussed in previous books in this series, the gaseous nature of oxygen and nitrogen in ambient conditions is perfectly fit for air-breathing beings of our physiological design.

9. Atkins offers a geographical word picture to help convey the special relation of hydrogen to the rest of the periodic table. "North of the mainland, situated rather like Iceland off the northwestern edge of Europe, lies a single isolated region—hydrogen," he writes. "This simple but gifted element is an essential outpost of the kingdom, for despite its simplicity it is rich in chemical personality. It is also the most abundant element in the universe and the fuel of the stars" (Atkins, *The Periodic Kingdom*, 90).

10. Herein we are considering the most common C-H bond, between a Csp3 carbon and hydrogen. Chemists typically classify bonds between atoms that differ in electronegativity by less than 0.4 to be nonpolar, although some electron disbalancing still occurs and other types of C-H bonds may display moderate to high polarity.

11. And it's not just polar solutes and charged ions (such as Na^+ and Cl^-) which are attracted to the electronegative oxygen atoms or the electropositive hydrogen atoms. The atoms of the water molecules themselves form an electrostatic network (described as a hydrogen-bonded network) where the hydrogen atoms of one water molecule are attracted to the oxygen atoms of adjacent water molecules. In this way all the water molecules in a body of water become connected together in a vast web of electrostatic interactions.

12. J. P. Trinkaus, *Cells into Organs* (New Jersey: Prentice–Hall, 1984), 51.

13. Charles Tanford, "How Protein Chemists Learned about the Hydrophobic Factor: Protein Chemists and the Hydrophobic Factor," *Protein Science* 6, no. 6 (1997): 1365. [emphasis in original]

14. John Keats, "Ode on a Grecian Urn," 1820, available at *Poetry Foundation*, https://www.poetryfoundation.org/poems/44477/ode-on-a-grecian-urn.

15. Trinkaus, *Cells into Organs*, 52. [emphasis in original]

16. Trinkaus, *Cells into Organs*, 53.

17. Trinkaus, *Cells into Organs*, 53.

18. Trinkaus, *Cells into Organs*, 53.

19. Stephane Romero et al., "Filopodium Retraction Is Controlled by Adhesion to Its Tip," *Journal of Cell Science* 125, no. 21 (2012): 4999–5004, https://doi.org/10.1242/jcs.104778.

20. Herein we are considering the most common C-H bond, between a Csp3 carbon and hydrogen. Chemists typically classify bonds between atoms that differ in electronegativity by less than 0.4 to be nonpolar, although some electron disbalancing still occurs and other types of C-H bonds may display moderate to high polarity.

21. Alex L. Kolodkin and Marc Tessier-Lavigne, "Growth Cones and Axon Pathfinding," in *Fundamental Neuroscience*, 4th ed. (Boston: Elsevier, 2013), 363–384.

22. C. H. Waddington, *New Patterns in Genetics and Development* (New York: Columbia University Press, 1962), 105–107.

23. J. B. Edelmann and M. J. Denton, "The Uniqueness of Biological Self-Organization: Challenging the Darwinian Paradigm," *Biology & Philosophy* 22, no. 4 (2006): 589. https://doi.org/10.1007/s10539-006-9055-5. [internal references removed]

24. Wieland B. Huttner and Anne A. Schmidt, "Membrane Curvature: A Case of Endofelin," *Trends in Cell Biology* 12, no. 4 (2002): 155.

25. Harold P. Erickson, "Size and Shape of Protein Molecules at the Nanometer Level Determined by Sedimentation, Gel Filtration, and Electron Microscopy," *Biological Procedures Online* 11, no. 1 (2009): 32–51, https://doi.org/10.1007/s12575-009-9008-x.

26. Bruce Alberts et al., "Ion Channels and the Electrical Properties of Membranes," *Molecular Biology of the Cell*, 4th ed. (New York: Garland Science, 2002), https://www.ncbi.nlm.nih.gov/books/NBK26910/.

27. Bruce Alberts et al., "Cell–Cell Adhesion," *Molecular Biology of the Cell*, 4th ed. (New York: Garland Science, 2002), https://www.ncbi.nlm.nih.gov/books/NBK26937/.

28. David E. Green and Robert F. Goldberger, *Molecular Insights into the Living Process* (New York: Academic Press, 1967).

29. Arthur E. Needham, *The Uniqueness of Biological Materials* (London: Pergamon Press, 1965), 82, 90.

30. Robert R. Crichton, *Biological Inorganic Chemistry: A New Introduction to Molecular Structure and Function*, 2nd ed. (Amsterdam: Elsevier, 2012), 184.

31. Crichton, *Biological Inorganic Chemistry*, 187. See also Figure 9.12.

32. Alberts et al., "Ion Channels and the Electrical Properties of Membranes."

33. Alberts et al., "Ion Channels and the Electrical Properties of Membranes."

34. Alberts et al., "Ion Channels and the Electrical Properties of Membranes."

5. ENERGY FOR CELLS

1. Peter Atkins, *The Periodic Kingdom* (New York: Basic Books, 1995), 27.

2. Mark Twain, *Roughing It* (New York: Harper and Brothers, 1899), 259.

3. Paul Davies, "The 'Give Me a Job' Microbe," *The Wall Street Journal*, December 4, 2010.

4. Felisa Wolfe-Simon et al., "A Bacterium That Can Grow by Using Arsenic Instead of Phosphorus," *Science* 332, no. 6034 (June 3, 2011): 1163.

5. Alla Katsnelson, "Arsenic-Eating Microbe May Redefine Chemistry of Life," *Nature* (December 2, 2010).

6. "Arsenic-Loving Bacteria May Help in Hunt for Alien Life," BBC News, December 2, 2010, https://www.bbc.com/news/science-environment-11886943."

7. Erika Check Hayden, "Critics Weigh in on Arsenic Life," *Nature* (May 27, 2011), https://doi.org/10.1038/news.2011.333.

8. Tobias J. Erb et al., "GFAJ-1 Is an Arsenate-Resistant, Phosphate-Dependent Organism," *Science* 337, no. 6093 (July 27, 2012): 467–470, https://doi.org/10.1126/science.1218455.

9. F. Westheimer, "Why Nature Chose Phosphates," *Science* 235, no. 4793 (1987): 1173–1178.

10. Mostafa I. Fekry, Peter A. Tipton, and Kent S. Gates, "Kinetic Consequences of Replacing the Internucleotide Phosphorus Atoms in DNA with Arsenic," *ACS Chemical Biology* 6, no. 2 (2011): 127–130, https://doi.org/10.1021/cb2000023.

11. "Many tasks that a cell must perform, such as movement and the synthesis of macromolecules, require energy. A large portion of the cell's activities are therefore devoted to obtaining energy from the environment and using that energy to drive energy-requiring reactions. Although enzymes control the rates of virtually all chemical reactions within cells, the equilibrium position of chemical reactions is not affected by enzymatic catalysis. The laws of thermodynamics govern chemical equilibria and determine the energetically favorable direction of all chemical reactions. Many of the reactions that must take place within cells are energetically unfavorable, and are therefore able to proceed only at the cost of additional energy input. Consequently, cells must constantly expend energy derived from the environment." From Geoffrey M. Cooper, "Metabolic Energy," in *The Cell: A Molecular Approach*, 2nd ed. (Sunderland, MA: Sinauer Associates, 2000), https://www.ncbi.nlm.nih.gov/books/NBK9903/. The generation and use of metabolic energy is thus fundamental to all of cell biology, a point also emphasized in John Conrad Waterlow, Peter J. Garlick, and D. J. Millward, *Protein Turnover in Mammalian Tissues and in the Whole Body* (Amsterdam: Elsevier, Amsterdam, 1978), chap. 14.

12. Nick Lane, *The Vital Question: Why Is Life the Way It Is?*, 1st American ed. (New York: W. W. Norton, 2015), 64.

13. Lane, *The Vital Question*, 63.

14. R. K. Suarez, "Oxygen and the Upper Limits to Animal Design and Performance," *The Journal of Experimental Biology* 201, no. 8 (1998): 1065–1072. [internal references removed]

15. F. H. Westheimer, "Why Nature Chose Phosphates," *Science* 235 (1987): 1173. The "principal reservoirs of biochemical energy" that Westheimer speaks of are adenosine triphosphate (ATP), creatine phosphate, and phosphoenolpyruvate.

16. Westheimer, "Why Nature Chose Phosphates," 1176. [internal references removed]

17. Michael Denton, *Nature's Destiny: How the Laws of Biology Reveal Purpose in the Universe* (New York: The Free Press, 1998), 404–405. Interestingly, the phosphates have another key biological role which again indicates their remarkable stability in an aqueous medium. They link together the nucleotide bases in the DNA. As Westheimer points out, in constructing a "tape" from small molecules, the connecting groups must be at least divalent in order to supply one connection to each of the two adjacent nucleotides. Further, because the DNA helix functions in an aqueous environment, and as water tends to cause hydro-

lytic breakdown of ester bonds, it is also an advantage, perhaps even essential, that the connecting groups carry a negative charge. The phosphate groups in the DNA do in fact carry a negative charge, and this negative charge greatly retards the rate of hydrolysis of DNA. Westheimer, commenting on the role of phosphate in linking the nucleotides in DNA, remarks that any such compound must be at least divalent; it must also possess a negative charge, and the charge must be physically close to the two ester bonds to protect them against hydrolysis. "All of these conditions are met by phosphoric acid and no other alternative is obvious," he writes. (Westheimer, "Why Nature Chose Phosphates," 1177). No other compounds possess the correct mix of properties to drive the chemical machinery of the cell, or as Westheimer puts it, "No other residue appears to fulfil the multiple roles of phosphate in biochemistry" (Westheimer, "Why Nature Chose Phosphates," 1173).

18. Arthur Needham, *The Uniqueness of Biological Materials* (London: Pergamon Press, 1965), 404. [internal references removed]

19. Atkins, *The Periodic Kingdom*, 27.

20. Rob Phillips et al., *Physical Biology of the Cell*, 2nd edition (New York: Garland Science, 2013), 192.

21. Needham, *Uniqueness*, 403, quoting Edward O'Farrell Walsh, *An Introduction to Biochemistry* (London: English University Press, 1961).

22. Jeremy M. Berg, John L. Tymoczko, and Lubert Stryer, "Glycolysis Is an Energy-Conversion Pathway in Many Organisms," in *Biochemistry*, 5th ed. (New York: W. H. Freeman, 2002), https://www.ncbi.nlm.nih.gov/books/NBK22593/.

23. For more on the biochemistry of cellular respiration, see "Cellular Respiration: Or, How One Good Meal Provides Energy for the Work of 75 Trillion Cells," IUPUI Department of Biology (website), accessed April 27, 2020, https://www.biology.iupui.edu/biocourses/N100/2k4ch7respirationnotes.html.

24. George Wald, "The Origin of Life," *Scientific American* 191, no. 2 (1954): 53.

25. Peter Mitchell, "Coupling of Phosphorylation to Electron and Hydrogen Transfer by a Chemi-osmotic Type of Mechanism," *Nature* 191, no. 4784 (1961): 144–148. Also see Lane, *The Vital Question*, chap. 2.

26. Peter Mitchell won the Nobel Prize in Chemistry for his theory in 1978. See "The Nobel Prize in Chemistry 1978," The Nobel Prize (website), 2019, https://www.nobelprize.org/prizes/chemistry/1978/summary/.

27. Leslie Orgel, quoted in Lane, *The Vital Question*, 68.

28. "Bacteria use enormously diverse energy sources. Some, like animal cells, are aerobic; they synthesize ATP from sugars they oxidize to CO_2 and H_2O by glycolysis, the citric acid cycle, and a respiratory chain in their plasma membrane that is similar to the one in the inner mitochondrial membrane. Others are strict anaerobes, deriving their energy either from glycolysis alone (by fermentation) or from an electron-transport chain that employs a molecule other than oxygen as the final electron acceptor. The alternative electron acceptor can be a nitrogen compound (nitrate or nitrite), a sulfur compound (sulfate or sulfite), or a carbon compound (fumarate or carbonate), for example. The electrons are transferred to these acceptors by a series of electron carriers in the plasma membrane that are comparable to those in mitochondrial respiratory chains… Despite this diversity, the plasma membrane of the vast majority of bacteria contains an ATP synthase that is very similar to the one in mitochondria. In bacteria that use an electron-transport chain to harvest energy, the electron-transport pumps H^+ out of the cell and thereby establishes a proton-motive force

across the plasma membrane that drives the ATP synthase to make ATP. In other bacteria, the ATP synthase works in reverse, using the ATP produced by glycolysis to pump H^+ and establish a proton gradient across the plasma membrane. The ATP used for this process is generated by fermentation processes... Thus, most bacteria, including the strict anaerobes, maintain a proton gradient across their plasma membrane. It can be harnessed to drive a flagellar motor, and it is used to pump Na^+ out of the bacterium via a Na^+-H^+ antiporter that takes the place of the Na^+-K^+ pump of eucaryotic cells. This gradient is also used for the active inward transport of nutrients, such as most amino acids and many sugars: each nutrient is dragged into the cell along with one or more H^+ through a specific symporter (Figure 14-32). In animal cells, by contrast, most inward transport across the plasma membrane is driven by the Na^+ gradient that is established by the Na^+-K^+ pump." From Bruce Alberts et al., "Electron-Transport Chains and Their Proton Pumps," *Molecular Biology of the Cell*, 4th ed. (New York: Garland Science, 2002), available at https://www.ncbi.nlm.nih.gov/books/NBK26904/; see also Werner Kühlbrandt, "Structure and Function of Mitochondrial Membrane Protein Complexes," *BMC Biology* 13, no. 1 (December 2015).

29. Lane, *The Vital Question*, 13.

30. Alberts et al., *Molecular Biology of the Cell*.

31. Nick Lane, John F. Allen, and William Martin, "How Did LUCA Make a Living? Chemiosmosis in the Origin of Life," *BioEssays* 32, no. 4 (January 27, 2010): 271–280, https://doi.org/10.1002/bies.200900131.

32. Lane, *The Vital Question*, 68.

33. Lane, *The Vital Question*, 68, figure 8.

34. Lane, *The Vital Question*, 68.

35. Berg, Tymoczko, and Stryer, "A Proton Gradient Powers the Synthesis of ATP," in *Biochemistry*, https://www.ncbi.nlm.nih.gov/books/NBK22388/.

36. Lane, *The Vital Question*, 69–70.

37. Lane, *The Vital Question*, 70.

38. Lane, *The Vital Question*, 70–71. [emphasis original, internal references removed]

39. "Iron has the possibility of existing in various oxidation states [possessing variable numbers of electrons]... which can be fine-tuned by appropriate choice of ligands [compounds linked to iron atom by coordinate bond] to encompass almost the entire biologically significant range of redox potentials." From Robert R. Crichton, *Biological Inorganic Chemistry: A New Introduction to Molecular Structure and Function*, 2nd ed. (Amsterdam: Elsevier, 2012), 248.

40. R. J. P. Williams, "The Symbiosis of Metal and Protein Function," *European Journal of Biochemistry* 150 (1985): 245.

41. Williams, "The Symbiosis of Metal and Protein Function," 245–246.

42. J. J. R. Fraústo da Silva and R. J. P. Williams, *The Biological Chemistry of the Elements: The Inorganic Chemistry of Life*, 2nd ed. (New York: Oxford University Press, 1991), 107.

43. Crichton, *Biological Inorganic Chemistry*, 248.

6. No Biology Without Metals

1. A. J. Gurevich, *Categories of Medieval Culture* (London: Routledge and Kegan Paul, 1985), 57–59.

2. John W. Morgan and Edward Anders, "Chemical Composition of Earth, Venus, and Mercury," *PNAS* 77, no. 12 (December 1, 1980): 6973–6977, https://doi.org/10.1073/pnas.77.12.6973; https://chem.libretexts.org/Bookshelves/Environmental_Chemistry/Book%3A_Geochemistry_(Lower)/The_Earth_and_its_Lithosphere/The_Earth%27s_crust.

3. Robert J. P. Williams, "The Symbiosis of Metal and Protein Functions," *European Journal of Biochemistry* 150 (1985): 232, https://febs.onlinelibrary.wiley.com/doi/pdf/10.1111/j.1432-1033.1985.tb09013.x.

4. Williams, "The Symbiosis of Metal and Protein Functions," 247. Another excellent review of the role of metals in biology is given in J. J. R. Fraústo da Silva and R. J. P. Williams, *The Biological Chemistry of the Elements: The Inorganic Chemistry of Life*, 2nd ed. (Oxford: Oxford University Press, 1991).

5. Earl Frieden, "The Evolution of Metals as Essential Elements," in *Protein-Metal Interactions*, ed. M. Friedman (New York: Plenum Press, 1974), 11.

6. Todor Dudev and Carmay Lim, "Principles Governing Mg, Ca and Zn Binding and Selectivity in Proteins," *Chemical Review* 103 (2003): 773. [emphasis in original]

7. Williams, "The Symbiosis of Metal and Protein Functions," 243. Williams writes, "For example iron can be taken up entirely selectively as FeOFe (in hemerythrin and ribonuclease reductase), Fe_2S_2, Fe_3S_3, Fe_4S_4 (ferrodoxins), Fe (many oxidases)… Variation in the associated proteins with any one of these chemical forms of iron has generated a wide range of enzymes… In several cases the iron itself can be held in many redox states and many spin states and, most unlike the amino acids, it can be held in many stereochemistries…. an iron atom is not an entity like an amino-acid side-chain… but is so variable that functionally it can be as different as a histidine (acting as an acid) on the one hand, iron phosphatases, and a thiolate (acting as a redox agent) on the other, cytochrome *c*. Such a metal atom can not be considered as a single functional group but has a variety of potential functions."

8. Robert R. Crichton, *Biological Inorganic Chemistry: A New Introduction to Molecular Structure and Function*, 2nd ed. (Amsterdam: Elsevier, 2012), 3.

9. Crichton, *Biological Inorganic Chemistry*, xi.

10. Crichton, *Biological Inorganic Chemistry*, 2.

11. Crichton, *Biological Inorganic Chemistry*, 3.

12. Crichton, *Biological Inorganic Chemistry*, 3.

13. Crichton, *Biological Inorganic Chemistry*, 3.

14. Lawrence Henderson, *The Fitness of the Environment* (New York: MacMillan, 1913), 240, https://archive.org/details/cu31924003093659/page/n209.

15. Henderson, *Fitness*, 241.

16. Henderson, *Fitness*, 191.

17. Silva and Williams, *The Biological Chemistry of the Elements*; Wolfgang Kaim, Brigitte Schwederski, and Axel Klein, *Bioinorganic Chemistry: Inorganic Elements in the Chemistry of Life: An Introduction and Guide*, 2nd ed. (Chichester, West Sussex, UK: Wiley, 2013); Crichton, *Biological Inorganic Chemistry*. Even a cursory reading of any of these books reveals that Williams's claims have been massively confirmed over the past three decades.

18. Crichton, *Biological Inorganic Chemistry*, chap. 1; "Essential Elements for Life," LibreTexts (website), last modified June 16, 2019, https://chem.libretexts.org/Textbook_Maps/

General_Chemistry/Map%3A_Chemistry_(Averill_and_Eldredge)/01%3A_Introduction_to_Chemistry/1.8%3A_Essential_Elements_for_Life.

19. Crichton, *Biological Inorganic Chemistry*, 5.

20. Bruce Alberts et al., "Ion Channels and the Electrical Properties of Membranes," in *Molecular Biology of the Cell*, 4th ed. (New York: Garland Science, 2002), https://www.ncbi.nlm.nih.gov/books/NBK26910/.

21. In fact the calcium triggers muscle contraction by deactivating the proteins which inhibit the force-producing actin-myosin interaction.

22. Williams, "The Symbiosis of Metal and Protein Functions," 238.

23. J. A. Cowan, ed., *The Biological Chemistry of Magnesium* (London: J. Wiley and Sons, 1995). See Chapter 1. Calcium binds 1,000 times more strongly than magnesium. S. S. Rosenfeld and E. W. Taylor, "Kinetic Studies of Calcium and Magnesium Binding to Troponin C," *Journal of Biological Chemistry* 260 (1985): 242–251. For an explanation of why see Crichton, *Biological Inorganic Chemistry*, 198, 215. First, magnesium requires a more regular geometrical binding site than calcium, and such sites are difficult to arrange in a protein because of the basic irregularity of its structure; and secondly, proteins in their molecular irregularity and in their possession of readily accessible oxygen atoms provide an ideal molecular matrix for the design of calcium-binding sites. Crichton comments further: "Like Na^+, Mg^{2+} is invariably hexacoordinate, whereas both K^+ and Ca^{2+} can adjust easily to 6, 7 or 8 coordination. Thus, Ca^{2+} can accommodate a more flexible geometry, compared to the octahedral geometry of the obligatory hexacoordinate cations, resulting in deviations from the expected bond angle of 90° by up to 40° compared with less than half that for Mg^{2+}. Similarly, bond lengths for oxy-ligands can vary by as much as 0.5 Å for Ca^{2+} whereas the corresponding values for Mg^{2+} vary by only 0.2 Å" [198]. Another reason for the weaker binding of Mg^{++} is that because of its small ionic radius it tends to bind to smaller water molecules rather than to protein ligands. As Crichton comments, "Many Mg^{2+}-binding sites in proteins have only 3, 4 or even less direct binding contacts to the protein, leaving several sites… occupied by water" (198).

24. Williams, "The Symbiosis of Metal and Protein Functions."

25. Williams, "The Symbiosis of Metal and Protein Functions," 238.

26. Wilhelm Jahnen-Dechent and Markus Ketteler, "Magnesium Basics," *Clinical Kidney Journal* 5, Supplement 1 (2012): 13–14.

27. Crichton, *Biological Inorganic Chemistry*, chap. 10.

28. Cowan, *The Biological Chemistry of Magnesium*, chap. 1.

29. The properties of the magnesium atom that equip it uniquely for its various roles in the cell are described in many publications. Cowan, *The Biological Chemistry of Magnesium*, chap. 1; M. Susan Cates et al., "Molecular Mechanisms of Calcium and Magnesium Binding to Parvalbumin," *Biophysical Journal* 82, no. 3 (March 2002): 1133–1146, https://doi.org/10.1016/S0006-3495(02)75472-6. And see Crichton, *Biological Inorganic Chemistry*, chap. 10.

30. See Crichton, *Biological Inorganic Chemistry*, chap. 10; J. M. Berg, J. L. Tymoczko, and L. Stryer, "Nucleoside Monophosphate Kinases: Catalyzing Phosphoryl Group Exchange between Nucleotides Without Promoting Hydrolysis," in *Biochemistry*, 5th ed. (New York: W. H. Freeman, 2002), https://www.ncbi.nlm.nih.gov/books/NBK22514/#A1242; Till Rudack et al., "The Role of Magnesium for Geometry and Charge in GTP Hydrolysis,

Revealed by Quantum Mechanics/Molecular Mechanics Simulations," *Biophysical Journal* 103, no. 2 (2012): 293–302.

31. Crichton, *Biological Inorganic Chemistry*, chap. 10; Rudack et al., "The Role of Magnesium for Geometry and Charge."

32. Peter Atkins, *The Periodic Kingdom* (New York: Basic Books, 1995), 16.

33. Crichton, *Biological Inorganic Chemistry*, 111.

34. Kaim et al., *Bioinorganic Chemistry*, 59–67.

35. Melvin Calvin, "Evolutionary Possibilities for Photosynthesis and Quantum Conversion," in *Horizons in Biochemistry*, eds. Michael Kasha and Bernard Pullman (New York: Academic Press, 1962), 53. For discussion of the unique properties of Mg in chlorophyll, see also J. Katz, "Chlorophyll," in *Inorganic Biochemistry*, vol. 2, ed. G. L. Eichhorn (Amsterdam: Elsevier, 1973), 1022–1066, especially pages 1025–1026.

36. Williams, "The Symbiosis of Metal and Protein Functions," 241.

37. G. Feher et al., "Structure and Function of Bacterial Photosynthetic Reaction Centres," *Nature* 339 (1989): 111, https://doi.org/10.1038/339111a0.

38. Crichton, *Biological Inorganic Chemistry*, 5. See also B. Halliwell and J. M. Gutteridge, "Oxygen Toxicity, Oxygen Radicals, Transition Metals and Disease," *The Biochemical Journal* 219, no. 1 (1984): 1–14. As they point out, the electronic configuration of the oxygen molecules containing two unpaired electrons "imposes a restriction on oxidations by O_2 which tends to make O_2 accept its electrons one at a time and slows its reactions with non radical species. Transition metals are found at the active site of many oxidases and oxygenases because their ability to accept and donate single electrons can overcome this spin restriction." See also Kasper P. Jensen and Ulf Ryde, "How O_2 Binds to Heme: Reasons for Rapid Binding and Spin Inversion," *Journal of Biological Chemistry* 279, no. 15 (2004): 14561–14569. The authors comment on the relative inactivity of the O_2 molecule: "Nature has handled this problem by using transition metals to carry, activate, and reduce O_2. There are many reasons for this choice. First, most transition metals also contain unpaired electrons, allowing reactions with triplet O_2. Second, transition metals are relatively heavy atoms, which increases spin-orbit coupling (SOC), and thereby provide a quantum mechanical mechanism to change the spin state of an electron, called spin inversion. However, the SOC of the first-row transition metals is too small alone to allow for spin transitions. Third, transition metals often have several excited states with unpaired electrons close in energy to the ground state. This can also be used to enhance the probability of spin inversion." [internal references removed] For a popular account of the role of iron in activating oxygen see Chapter 6 in Nick Lane, *Oxygen: The Molecule That Made the World* (Oxford: Oxford University Press, 2002).

39. See Kaim et al., *Bioinorganic Chemistry*, 82–91. They write, "Certain groups of mollusks, crustaceans, spiders and worms on one hand and the majority of other organisms, particularly vertebrates, on the other differ in their strategies in O_2 coordination. While the former contain dinuclear [transitional] metal arrangements with amino acid coordination, namely the copper protein hemocyanin or the iron protein haemerythrin, most other breathing organisms use the heme system, that is, monoiron complexes of a certain porphyrin macrocycle protoporphyrin IX." [internal references removed]

40. Kaim et al., *Bioinorganic Chemistry*, 90.

41. J. Mitchell Salhany, "Effect of Carbon Dioxide on Human Hemoglobin: Kinetic Basis for the Reduced Oxygen Affinity," *Journal of Biological Chemistry* 247 (June 25, 1972): 3799–3801.

42. "O_2 is not a very good ligand for this system. Other ligands such as CO and CN- can bind more strongly to the iron because they are better pi acceptors than oxygen, and the fact that they can block the O_2 binding site so effectively is what makes them so very toxic. However, it is critical for the body that O_2 be only weakly bound because it needs to be able to pop on and off of the iron where it is needed. In other words, the reversibility of the reaction is important. While O_2 has two electrons in its pi* orbital, CO and CN- have lots of empty, electron-accepting pi* space that strengthens their bonds to iron. Note that although CO and CN- are isoelectronic (meaning that their MO configuration is the same) the negative charge on CN makes it a slightly worse electron acceptor than CO and therefore CO binds more strongly to the iron than CN-. Paradoxically, CN- is more toxic per mole than CO. This is probably because it catalyzes some other configuration change in the protein chain that permanently deactivates the heme unit." From "Hemoglobin," Stanford University (website), accessed May 13, 2019, https://web.stanford.edu/~kaleeg/chem32/hemo/.

43. Williams, "The Symbiosis of Metal and Protein Function."

44. Halliwell and Gutteridge, "Oxygen Toxicity, Oxygen Radicals, Transition Metals and Disease."

45. Jensen and Ryde, "How O_2 Binds to Heme."

46. Crichton, *Biological Inorganic Chemistry*, 254. See also Kaim et al., *Bioinorganic Chemistry*, 88.

47. Crichton, *Biological Inorganic Chemistry*, 248.

48. Kaim et al., *Bioinorganic Chemistry*, 89.

49. "We show that porphyrin is an ideal iron ligand for the spin transition problem, because it tunes the spin states to be close in energy, giving parallel binding curves, small activation energies, and large transition probabilities. This finding explains why the porphyrin ring is designed to bring spin states close in energy and why spin inversion and reversible binding is possible in heme proteins. We also provide evidence that similar arguments apply to other heme proteins, *e.g.* the heme peroxidases, where near degeneracy, in this case in the ferric state, is caused by strengthening the ligand field of the proximal histidine by a hydrogen bond to a carboxylate group. Hence, we suggest a new role for the choice of an axial ligand in such systems, *viz.* to bring spin states close in energy and thereby facilitate spin-forbidden binding of ligands." Jensen and Ryde, "How O_2 Binds to Heme," 14562.

50. Frieden, "Evolution of Metals," 22. [emphasis in original]

51. Robert Gennis and Shelagh Ferguson-Miller, "Structure of Cytochrome c Oxidase, Energy Generator of Aerobic Life," *Science* 269 (1995): 1063–1064, https://doi.org/10.1126/science.7652553. And see Figure 14-27 at Alberts et al., *Molecular Biology of the Cell*.

52. Antony Crofts, "Cytochrome Oxidase," University of Illinois at Urbana-Champaign (website), 1996, https://www.life.illinois.edu/crofts/bioph354/cyt_ox.html.

53. Darryl Horn and Antoni Barrientos, "Mitochondrial Copper Metabolism and Delivery to Cytochrome C Oxidase," *IUBMB Life* 60, no. 7 (July 2008): 421–429, https://doi.org/10.1002/iub.50.

54. Crichton, *Biological Inorganic Chemistry*, 312.

55. Crichton, *Biological Inorganic Chemistry*, 314.

56. Lane, *Oxygen*, 145.

57. Crichton, *Biological Inorganic Chemistry*, 12.

58. Crichton, *Biological Inorganic Chemistry*, 12.

59. Crichton, *Biological Inorganic Chemistry*, 229–230. The unique electronic properties of the divalent zinc ion allow zinc to adopt highly flexible coordination geometry, though "in most zinc proteins," explains Crichton, "there is a strong preference for tetrahedral coordination." (Tetrahedral coordination involves bonds to four other atoms.)

60. For more about carbonic anhydrase, see the collection of resources at "Carbonic Anhydrase," ScienceDirect, accessed May 7, 2020,

https://www.sciencedirect.com/topics/materials-science/carbonic-anhydrase.

61. Haewon Park, Patrick J. McGinn, and François M. M. Morel, "Expression of Cadmium Carbonic Anhydrase of Diatoms in Seawater," *Aquatic Microbial Ecology* 51 (2008): 183–193.

62. Estimated from oxygen utilization per second given in John N. Maina, "Comparative Respiratory Morphology: Themes and Principles in the Design and Construction of the Gas Exchangers," *The Anatomical Record* 261, no. 1 (2000): 26.

63. Jane Higdon, "Molybdenum," Linus Pauling Institute: Micronutrient Information Center, Oregon State University, 2001, accessed July 19, 2019, https://lpi.oregonstate.edu/mic/minerals/molybdenum.

64. Crichton, *Biological Inorganic Chemistry*, 335.

65. Crichton, *Biological Inorganic Chemistry*, 337.

66. Crichton, *Biological Inorganic Chemistry*, 337.

67. L. J. Rothschild and R. L. Mancinelli, "Life in Extreme Environments," *Nature* 409, no. 6823 (2001): 1092–1101.

68. Williams, "The Symbiosis of Metal and Protein Function," 247.

7. THE MATRIX

1. Anders Nilsson and Lars G. M. Pettersson, "The Structural Origin of Anomalous Properties of Liquid Water," *Nature Communications* 6 (December 8, 2015): 8998, https://doi.org/10.1038/ncomms9998.

2. David Rogers, "Neutrophil Chasing Bacteria," Embryology Education and Research, video, 0:33, accessed May 18, 2020, https://embryology.med.unsw.edu.au/embryology/index.php/Movie_-_Neutrophil_chasing_bacteria.

3. Albert Szent-Gyorgyi, *Bioelectronics* (New York: Academic Press, 1968), 9.

4. Water possesses a host of additional properties beneficial for life on Earth, ones not directly relevant to the functioning of the cell. Its density decreases below 4 C and expands on freezing, which prevents it from freezing from the bottom up, preserving life in oceans, lakes, and rivers; it exists in the three material states—vapor, liquid, and solid—in the ambient temperature range, enabling the vital hydrological cycle, which delivers water to land-based life… and so forth. For many more such instances, see Michael Denton, *The Wonder of Water: Water's Profound Fitness for Life on Earth and Mankind* (Seattle, WA: Discovery Institute Press, 2017).

5. See "Molecule Diffusion," European Advanced Light Microscopy Network (website), February 5, 2004, https://www.embl.de/eamnet/frap/html/molecule_diffusion.html.

6. Glenn Elert, "Viscosity," The Physics Hypertextbook, accessed May 14, 2019, https://physics.info/viscosity/.

7. See, for example, Bruce Alberts et al., "The Self-Assembly and Dynamic Structure of Cytoskeletal Filaments," *Molecular Biology of the Cell*, 4th ed. (New York: Garland Science, 2002), https://www.ncbi.nlm.nih.gov/books/NBK26862/; Adam J. Kuskak, "How Cells Crawl: Illuminating the Dynamics of Cell Motility," *The NIH Catalyst* 21, no. 5 (2013), https://irp.nih.gov/catalyst/v21i5/how-cells-crawl; and Daniel A. Fletcher and R. Dyche Mullins, "Cell Mechanics and the Cytoskeleton," *Nature* 463 (January 27, 2010): 485–492, https://doi.org/10.1038/nature08908. The ability of embryonic cells to crawl depends on the existence within the cell of a network of molecular components which make up the cytoskeleton, including actin and intermediate fibers, microtubules, and various molecular motors such as myosin, which can impose contractile forces on various parts of the network. As described in "The Self-Assembly and Dynamic Structure of Cytoskeletal Filaments," "Three types of cytoskeletal filaments are common to many eukaryotic cells and are fundamental to the spatial organization of these cells. *Intermediate filaments* provide mechanical strength and resistance to shear stress. *Microtubules* determine the positions of membrane-enclosed organelles and direct intracellular transport. *Actin filaments* determine the shape of the cell's surface and are necessary for whole-cell locomotion... But these cytoskeletal filaments would be ineffective on their own. Their usefulness to the cell depends on a large number of accessory proteins that link the filaments to other cell components, as well as to each other. This set of *accessory proteins* is essential for the controlled assembly of the cytoskeletal filaments in particular locations, and it includes the *motor proteins* that either move organelles along the filaments or move the filaments themselves." [emphasis in original] During crawling, the activities of these components working mechanically together and linked to focal adhesions between the cell and the substratum push the cell forward. The process is complex and a number of models have been proposed. See A. D. Bershadsky and M. M. Kozlov, "Crawling Cell Locomotion Revisited," *PNAS* 108, no. 51 (December 20, 2011): 20275–20276, https://doi.org/10.1073/pnas.1116814108. But basically all models involve polymerization of actin fibers at the leading edge of the cell, which push the cell membrane forward. This is accompanied by the formation of new forward focal adhesions and the breaking of old ones towards the rear of the cell. As described by Revathi Ananthakrishnan and Allen Ehrlicher, "Cell movement is a complex phenomenon primarily driven by the actin network beneath the cell membrane, and can be divided into three general components: protrusion of the leading edge of the cell, adhesion of the leading edge and deadhesion at the cell body and rear, and cytoskeletal contraction to pull the cell forward. Each of these steps is driven by physical forces generated by unique segments of the cytoskeleton." "The Forces behind Cell Movement," *International Journal of Biological Sciences* 3, no. 5 (June 1, 2007): 303–17.

8. Rob Phillips et al., *Physical Biology of the Cell*, 2nd ed. (New York: Garland Science, 2013), 42.

9. Estimated from data at Ron Milo and Rob Phillips, "How Many Proteins are in a Cell?," Cell Biology by the Numbers (website), accessed May 13, 2019, http://book.bionumbers.org/how-many-proteins-are-in-a-cell/.

10. Estimated from data at Milo and Phillips, "What are the Concentrations of Cytoskeletal Molecules?," Cell Biology by the Numbers (website), accessed May 13, 2019, http://book.bionumbers.org/what-are-the-concentrations-of-cytoskeletal-molecules/.

11. Knut Schmidt-Nielsen, *Animal Physiology: Adaptation and Environment*, 5th ed. (Cambridge, UK: Cambridge University Press, 1997), Appendix B, Diffusion.

12. Stephane Romero et al., "Filopodium Retraction Is Controlled by Adhesion to Its Tip," *Journal of Cell Science* 125, no. 21 (November 1, 2012): 4999–5004.

13. Alex L. Kolodkin and Marc Tessier-Lavigne, "Growth Cones and Axon Pathfinding," *Fundamental Neuroscience* (Burlington, MA: Academic Press, 2013), 363–384.

14. Schmidt-Nielsen, *Animal Physiology*, chap. 2.

15. See "Distance between Capillaries in Adult Heart," BioNumbers, https://bionumbers.hms.harvard.edu/bionumber.aspx?s=n&v=3&id=113202; and Robert A. Freitas Jr., "8.2.1.2 Arteriovenous Microcirculation," *Nanomedicine, Volume I: Basic Capabilities* (Georgetown, TX: Landes Bioscience, 1999), http://www.nanomedicine.com/NMI/8.2.1.2.htm.

16. Phillips et al., *The Physical Biology of the Cell*, 497.

17. Steven Vogel, *Comparative Biomechanics: Life's Physical World*, 2nd ed. (Princeton: Princeton University Press, 2013), 187.

18. Dongdong Jia et al., "The Time, Size, Viscosity, and Temperature Dependence of the Brownian Motion of Polystyrene Microspheres," *American Journal of Physics* 75, no. 2 (February 2007): 111–115, https://doi.org/10.1119/1.2386163.

19. For the requirement for a liquid matrix (water in the case of Terran life) see N. V. Sidgwick, "Molecules," *Science* 86 (1937): 335–340; J. A. Baross et al., *The Limits of Organic Life in Planetary Systems* (Washington, DC: National Academies Press, 2007), 1; Kevin W. Plaxco and Michael Gross, *Astrobiology: A Brief Introduction*, 2nd ed. (Baltimore: Johns Hopkins University Press, 2011), 14; Arthur Needham, *The Uniqueness of Biological Materials* (Oxford: Pergamon Press, 1965), 9; Louis Irwin, *Cosmic Biology: How Life Could Evolve on Other Worlds* (New York: Springer and Praxis, 2011), 43, 303. The latter, for example, find the necessity for a liquid matrix "compelling" and liquids they claim provide "overwhelming advantages" over solids and gases to serve as the matrix of life.

20. See Needham, *Uniqueness*, 9.

21. Glenn Elert, "Viscosity," The Physics Hypertextbook, accessed May 14, 2019, https://physics.info/viscosity/.

22. Elert, "Viscosity."

23. Yaolin Shi and Jianling Cao, "Lithosphere Effective Viscosity of Continental China," *Earth Science Frontiers* 15, no. 3 (May 2008): 82–95, https://doi.org/10.1016/S1872-5791(08)60064-0.

24. Water's power as a solvent plays a vital role in phenomena well beyond the cell. Its assists in the weathering and erosion of the rocks distributing the key elements of life to terrestrial ecosystems. It enables life in the oceans, rivers, and lakes. It carries solutes and various nutrients through the micropores in the soil, nourishing plants and trees. It assists the transport up the stems and trunks of plants and trees to the leaves where the sun's energy is captured in photosynthesis, providing foodstuffs for terrestrial animal life. For more on the subject, see Michael Denton, *The Wonder of Water*.

25. Alok Jha, *The Water Book* (London: Headline, 2015), 24.

26. F. Franks, "Water the Unique Chemical," in *Water: A Comprehensive Treatise*, vol. 1 (New York: Plenum Press, 1972), 20.

27. Lawrence Henderson, *The Fitness of the Environment* (New York: MacMillan, 1913), 111, https://archive.org/details/cu31924003093659/page/n209.

28. The only polar (or charged) molecules which do not dissolve are very large molecules like cotton or cellulose.

29. Albert Szent-Gyorgyi, *The Living State: With Observations on Cancer* (New York: Academic Press, 1972), 9.

30. Philip Ball, "Water as an Active Constituent in Cell Biology," *Chemical Reviews* 108, no. 1 (2008): 74.

31. Ball, "Water as an Active Constituent," 100.

32. Ball, "Water as an Active Constituent," 103. [emphasis in original]

33. Gerald H. Pollack in *Water and the Cell*, ed. Gerald H. Pollack, Ivan L. Cameron, and Denys N. Wheatley (Dordrecht: Springer, 2006), viii.

34. Just to put a number on it, the bibliography section of chemist Martin Chaplin's Water Structure and Water website lists close to 4,000 references on water as of June 24, 2020. http://www1.lsbu.ac.uk/water/ref40.html

35. Water molecules form hydrogen bonds with neighboring water molecules. Each oxygen atom, which is negatively charged, forms two hydrogen bonds with the hydrogen atoms on neighboring water molecules. And this causes the formation of a network which extends right through the water.

36. Jha, *The Water Book*, 115–116. Antony Crofts provides additional information: "In ice, the H-bonded networks are more extensive, and ice is a better conductor than liquid water. Conduction occurs through a 'hop-turn' mechanism, first suggested by Grotthuss, and often referred to as the Grotthuss mechanism... In the 'hop' part of the mechanism, a proton first hops from the end of the H-bonded chain to an adjacent group (I, right); transfer of H-bond strength then allows it to be replaced by a H^+ binding at the other end, to give the structure in II. In the 'turn' phase, rotation of the waters as shown in II then restores the starting structure (I). In this H-bonded chain, the waters can in principle be replaced by suitable protein side chains with H-bonding potential." Antony Crofts, "Lecture 12: Proton Conduction, Stoichiometry" (lecture, University of Illinois at Urbana-Champaign, 1996), http://www.life.illinois.edu/crofts/bioph354/lect12.html.

37. Harold Morowitz, *Cosmic Joy and Local Pain* (New York: Scribner, 1987), 152. He adds that "proton conductance has become a subject of central interest in biochemistry because of its role in photosynthesis and oxidative phosphorylation" (153). As Morowitz explains, both these key processes use proton conductance and hydrated ions, which are major features of water. "Once again the fitness enters in, in the detailed way in which the molecular properties of water are matched to the molecular mechanisms of bio energetics" (154).

38. Nick Lane, *The Vital Question: Why Is Life the Way It Is?*, (New York: W. W. Norton & Company, 2015), 120.

39. Lane, *The Vital Question*, chap. 3.

40. L. J. Rothschild and R. L. Mancinelli, "Life in Extreme Environments," *Nature* 409, no. 6823 (February 22, 2001): 1092–1101; Juliette Ravaux et al., "Thermal Limit for Metazoan Life in Question: *In Vivo* Heat Tolerance of the Pompeii Worm," *PloS One* 8, no. 5 (2013): e64074, https://doi.org/10.1371/journal.pone.0064074.

41. Rothschild and Mancinelli, "Life in Extreme Environments."

42. W. J. Gehring and R. Wehner, "Heat Shock Protein Synthesis and Thermotolerance in Cataglyphis, an Ant from the Sahara Desert," *PNAS* 92, no. 7 (March 28, 1995): 2994–2998.

43. Ken Takai et al., "Cell Proliferation at 122°C and Isotopically Heavy CH4 Production by a Hyperthermophilic Methanogen under High-Pressure Cultivation," *PNAS* 105, no. 31 (August 5, 2008): 10949–10954, https://doi.org/10.1073/pnas.0712334105.

44. Bruce M. Jakosky et al., "Subfreezing Activity of Microorganisms and the Potential Habitability of Mars' Polar Regions," *Astrobiology* 3, no. 2 (June 2003): 343–350, https://doi.org/10.1089/153110703769016433.

45. Rothscild and Mancinelli, "Life in Extreme Environments."

46. Andrew Clarke et al., "A Low Temperature Limit for Life on Earth," *PloS One* 8, no. 6 (2013): e66207.

47. Clarke et al., "A Low Temperature Limit."

48. As I pointed out, "Even a temperature change of far less than 100°C causes a quite dramatic slowing of reaction times. Reactions occurring in the human body at 38°C would take place sixteen times slower at 0°C and sixty-four times slower at -20°C. As Robert E. D. Clark points out, at temperatures below -100°C all chemical reactions become vanishingly slow and at the temperature of liquid air, 'only a few reactions take place at all and these involve the exceedingly active element fluorine in its free state.' Even though some organic chemistry would be possible at temperatures as low as -40°C it would be… unimaginably slow." From Michael Denton, *Nature's Destiny* (London: The Free Press, 1998), 111. [internal references removed]; and see Robert E. D. Clark, *The Universe: Plan or Accident*, 3rd ed. (Grand Rapids: Zondervan, 1972), 98.

49. Water at a temperature of well over 100°C does exist under pressure in subsurface rocks, at the bottom of lakes in hot spring areas like Yellowstone and close to hydrothermal vents where the temperature of water is close to 400°C. Lake bottom water reaches 171°C. See Brett French, "Temperature from Yellowstone Lake Vents Hit New High," *Billings Gazette*, October 27, 2016, https://billingsgazette.com/lifestyles/recreation/temperatures-from-yellowstone-lake-vents-hit-new-high/article_4d088c9c-35ff-5041-a2f7-dc75dbdf5679.html.

50. Denton, *Nature's Destiny*, 114. See also J. Walker, "The Physics and Chemistry of the Lemon Meringue Pie," *Scientific American* 244, no. 6 (1981): 154–159, esp. pages 154–155.

51. G. N. Somero, "Proteins and Temperature," *Annual Review of Physiology* 57 (1995): 61.

52. Fred Hoyle, "Ultrahigh Temperatures," *Scientific American* 191, no. 3 (1954): 144–156.

53. Simon Mitton, *Cambridge Encyclopaedia of Astronomy* (London: Jonathan Cape, 1977), 128; Dedra Forbes, "Temperature at the Center of the Sun," The Physics Factbook (website), ed. Glenn Elert, accessed May 14, 2019, http://hypertextbook.com/facts/1997/DedraForbes.shtml.

54. R. Nave, "Stellar Spectral Types," Hyperphysics, accessed May 14, 2019, http://hyperphysics.phy-astr.gsu.edu/hbase/Starlog/staspe.html/.

55. Henderson, *Fitness*, 132.

56. Plaxco and Gross, *Astrobiology: A Brief Introduction*, 14–18. See also John T. Edsall and Jeffries Wyman, *Biophysical Chemistry: Thermodynamics, Electrostatics, and the Biological Significance of the Properties of Matter* (New York: Academic Press, 1958), 14, 17.

8. THE PRIMAL BLUEPRINT

1. Francis Crick, *Life Itself: Its Origin and Nature* (New York: Simon and Schuster, 1981), 88.

2. "Murchison," *The Meteoritical Bulletin Database*, The Meteoritical Society, May 11, 2019, accessed May 14, 2019, http://www.lpi.usra.edu/meteor/metbull.php?code=16875.

3. Keith Kvenvolden et al., "Evidence for Extraterrestrial Amino-Acids and Hydrocarbons in the Murchison Meteorite," *Nature* 228 (December 5, 1970): 923–926.

4. Philippe Schmitt-Kopplin et al., "High Molecular Diversity of Extraterrestrial Organic Matter in Murchison Meteorite Revealed 40 Years after Its Fall," *PNAS* 107, no. 7 (February 16, 2010): 2763–2768; Sandra Pizzarello et al., "Processing of Meteoritic Organic Materials as a Possible Analog of Early Molecular Evolution in Planetary Environments," *PNAS* 110, no. 39 (September 24, 2013): 15614–15619.

5. Glycine, alanine, and glutamic acid have been identified in the Murchison meteorite. See Kvenvolden et al., "Evidence for Extraterrestrial Amino-Acids."

6. Michael P. Callahan et al., "Carbonaceous Meteorites Contain a Wide Range of Extraterrestrial Nucleobases," *PNAS* 108, no. 34 (August 23, 2011): 13995–13998; Zita Martins et al., "Extraterrestrial Nucleobases in the Murchison Meteorite," *Earth and Planetary Science Letters* 270, no. 1–2 (June 2008): 130–136; Pascale Ehrenfreund and Jan Cami, "Cosmic Carbon Chemistry: From the Interstellar Medium to the Early Earth," *Cold Spring Harbor Perspectives in Biology* 2, no. 12 (December 1, 2010): a002097–a002097.

7. Although an impressive inventory of basic building blocks including several amino acids, nucleic acid bases, and sugars have been synthesized abiotically, there are still several key organic building blocks, including the porphyrins and the amino acids lysine and arginine, for which no plausible prebiotic synthesis has been achieved.

8. Sun Kwok and Yong Zhang, "Mixed Aromatic-Aliphatic Organic Nanoparticles as Carriers of Unidentified Infrared Emission Features," *Nature* 479, no. 7371 (November 3, 2011): 80–83.

9. Don McNaughton et al., "FT-MW and Millimeter Wave Spectroscopy of PANHs: Phenanthridine, Acridine, and 1,10-Phenanthroline," *The Astrophysical Journal* 678, no. 1 (May 2008): 309–315, doi:10.1086/529430.

10. Stanley L. Miller, "A Production of Amino Acids under Possible Primitive Earth Conditions," *Science* 117, no. 3046 (May 15, 1953): 528–529.

11. Eric T. Parker et al., "Primordial Synthesis of Amines and Amino Acids in a 1958 Miller H_2S-Rich Spark Discharge Experiment," *PNAS* 108, no. 14 (April 5, 2011): 5526–5531.

12. For an interactive graph of the abundances, see "Abundance in the Universe of the Elements," https://periodictable.com/Properties/A/UniverseAbundance.html.

13. Anne Marie Helmenstine, "Chemical Composition of the Human Body," ThoughtCo., February 11, 2019, https://www.thoughtco.com/chemical-composition-of-the-human-body-603995.

14. Robert R. Crichton, *Biological Inorganic Chemistry: A New Introduction to Molecular Structure and Function*, 2nd ed. (Oxford: Elsevier Science, 2012), 4.

15. Abigail C. Allwood et al., "Controls on Development and Diversity of Early Archean Stromatolites," *PNAS* 106, no. 24 (June 16, 2009): 9548–9555.

16. Guillermo Gonzalez, "What Astrobiology Teaches about the Origin of Life," ch. 15 of *The Mystery of Life's Origin: The Continuing Controversy* (Seattle: Discovery, 2020), 377.

17. D. P. Bartel, "Micro RNAs," *Cell* 126 (2009): 215–233.

18. Bruce Alberts et al., *Molecular Biology of the Cell*, 4th ed. (New York: Garland Science, 2002), https://www.ncbi.nlm.nih.gov/books/NBK21054/.

19. Addy Pross and Robert Pascal, "The Origin of Life: What We Know, What We Can Know and What We Will Never Know," *Open Biology* 3, (February 11, 2013): 1–5, https://doi.org/10.1098/rsob.120190; James D. Stephenson et al., "Boron Enrichment in Martian Clay," *PloS One* 8, no. 6 (2013): e64624; Eugene V. Koonin, "The Origins of Cellular Life," *Antonie van Leeuwenhoek* 106 (April 23, 2014): 27–41; Eugene V. Koonin and Artem S. Novozhilov, "Origin and Evolution of the Genetic Code: The Universal Enigma," *IUBMB Life* 61, no. 2 (February 2009): 99–111, https://doi.org/10.1002/iub.146; Jimmy Gollihar, Matthew Levy, and Andrew D. Ellington, "Many Paths to the Origin of Life," *Science* 343, no. 6168 (January 17, 2014): 259–260; Jan Spitzer, "Emergence of Life from Multicomponent Mixtures of Chemicals: The Case for Experiments with Cycling Physicochemical Gradients," *Astrobiology* 13, no. 4 (April 2013): 404–413; Sara Imari Walker, P. C. W. Davies, and George F. R. Ellis, eds., *From Matter to Life: Information and Causality* (Cambridge, UK: Cambridge University Press, 2017).

20. Stephen C. Meyer, *Signature in the Cell: DNA and the Evidence for Intelligent Design*, 1st ed. (New York: HarperOne, 2009).

21. Michael Denton, *Evolution: A Theory in Crisis*, 1st US ed. (Bethesda, MD: Adler & Adler, 1986).

22. Brian Miller, "Thermodynamic Challenges to the Origin of Life," in *The Mystery of Life's Origin: The Continuing Controversy* (Seattle: Discovery, 2020), 359–374.

23. Koonin and Novozhilov, "Origin and Evolution of the Genetic Code." [internal references removed]

24. Koonin and Novozhilov, "Origin and Evolution of the Genetic Code."

25. Koonin and Novozhilov, "Origin and Evolution of the Genetic Code." [internal references removed]

26. Koonin and Novozhilov, "Origin and Evolution of the Genetic Code."

27. "Prebiotic chemistry could produce a wealth of biomolecules from nonliving precursors. But the wealth soon became overwhelming, with the 'prebiotic soups' having the chemical complexity of asphalt (useful, perhaps, for paving roads but not particularly promising as a wellspring for life). Classical prebiotic chemistry not only failed to constrain the contents of the prebiotic soup, but also raised a new paradox: How would life (or any organized chemical process) emerge from such a mess?" *in* Steven A. Benner, "Origins of Life: Old Views of Ancient Events," *Science* 283, no. 5410 (1999): 2026. See also A. G. Cairns-Smith, *Genetic Takeover and the Mineral Origins of Life* (Cambridge, UK: Cambridge University Press, 1982); Charles B. Thaxton, Walter L. Bradley, and Roger L. Olsen, *The Mystery of Life's Origin: Reassessing Current Theories* (New York: The Philosophical Library of New York, 1984); Robert Shapiro, *Origins: A Skeptic's Guide to the Creation of Life on Earth* (New York: Summit Books, 1986); Gerald F. Joyce, "The Antiquity of RNA-Based Evolution," *Nature* 418, no. 6894 (2002): 214–221.

28. Joyce, "The Antiquity of RNA-Based Evolution," 215.

29. Joyce, "The Antiquity of RNA-Based Evolution," 215, Figure 2 caption.

30. Joyce, "The Antiquity of RNA-Based Evolution," 215.

31. Joyce, "The Antiquity of RNA-Based Evolution," 215. [internal references removed]

32. Shapiro, *Origins: A Skeptic's Guide*, 207.

33. Itay Budin and Jack W. Szostak, "Expanding Roles for Diverse Physical Phenomena During the Origin of Life," *Annual Review of Biophysics* 39 (2010): 245.

34. Paul Davies, *The Fifth Miracle: The Search for the Origin and Meaning of Life* (New York: Simon & Schuster, 1999), 260.

35. Davies, *The Fifth Miracle*, 261.

36. Davies, *The Fifth Miracle*, 263.

37. Tommaso Bellini et al., "Origin of Life Scenarios: Between Fantastic Luck and Marvelous Fine-Tuning," *Euresis* 2 (2012): 130.

38. Lawrence Henderson, *The Fitness of the Environment* (New York: MacMillan, 1913), 139. Available online at https://archive.org/details/cu31924003093659/page/n209.

39. Henderson, *Fitness*, 163.

40. Henderson, *Fitness*, 253.

41. Henderson, *Fitness*, 243–248.

42. Henderson, *Fitness*, 248.

43. Henderson, *Fitness*, 272.

44. Henderson, *Fitness*, 263.

45. Henderson, *Fitness*, 272.

46. Frances Westall and André Brack, "The Importance of Water for Life," *Space Science Reviews* 214 (2018): 7, https://doi.org/10.1007/s11214-018-0476-7.

47. Christopher P. McKay, "What is Life—And When Do We Search for It on Other Worlds," *Astrobiology* 20 (2020): 164, https://doi.org/10.1089/ast.2019.2136.

48. Aron Gurevich, *Categories of Medieval Culture* (London: Routledge and Kegan Paul, 1985), 59.

FIGURE CREDITS

Chapter 1

Figure 1.1. Radiolarian shells. "Haeckel Cyrtoidea." Plate 31 in Ernst Haeckel, *Kunstformen der Natur*, 1904. Public domain.

Chapter 2

Figure 2.1. Carbon. "Eight Allotropes of Carbon." Image by Michael Ströck (Mstroek), 2006, Wikimedia Commons. CC-BY-SA 3.0 license.

Figure 2.2. Cosmic temperatures. Image by Michael Denton and Brian Gage.

Chapter 3

Figure 3.1. DNA's double helix. Image by Brian Gage.

Figure 3.2. Covalent bonds in methane. "Covalent." Image by Dynablast, 2010, Wikimedia Commons. CC-BY-SA 2.5 license.

Figure 3.3. Energy levels. Image by Michael Denton and Brian Gage.

Chapter 4

Figure 4.1. Periodic table. "White Periodic Table." Image by Cepheus, 2007, Wikimedia Commons. Modified by Inverse Hypercube, 2012. Public domain.

Figure 4.2. The lipid bilayer cell membrane. "Cell Membrane Detailed Diagram." Image by Mariana Ruiz Villarreal (Lady of Hats), 2007, Wikimedia Commons. Public domain.

Chapter 5

Figure 5.1. Lake Mono, California. "Mono Lake South Tufa August 2013." Photograph by Tony (King of Hearts), 2013, Wikimedia Commons. CC-BY-SA 3.0 license.

Figure 5.2. Adenosine triphosphate (ATP). "ATPanionChemDraw." Image by Smokefoot, 2017, Wikimedia Commons. Public domain.

Figure 5.3. Ancient Roman drainage system at Rio Tito mines, Andalusia, Spain. "WaterwheelsSp." Image scan by Peterlewis of 1928 print, 2008, Wikimedia Commons. Public domain.

Chapter 6

Figure 6.1. Crab Nebula. "M1: The Crab Nebula from Hubble." Image by NASA/ESA, J. Hester and A. Loll. Used as permitted.

Figure 6.2. Rust on chain links near the Golden Gate Bridge in San Francisco. "RustChain." Photograph by Marlith, 2008, Wikimedia Commons. CC-BY-SA 3.0 license.

Figure 6.3. Rendering of human carbonic anhydrase II. "Carbonic Anhydrase 1CA2 Active Site." Image by Fvanesoncellos, 2011, Wikimedia Commons. Public domain.

Figure 6.4. MoFe cofactor in the MoFe protein. "FeMoCo Cluster." Image by Smokefoot, 2014, Wikimedia Commons. CC-BY-SA 4.0 license.

Chapter 7

Figure 7.1. Fetus in the womb. Illustration by Leonardo da Vinci, circa 1510. Public domain.

Figure 7.2. Water drop on leaf. "Water Drop on a Leaf." Photograph by Tanakawho, 2006, Wikimedia Commons. CC-BY 2.0 license.

Chapter 8

Figure 8.1. Cosmic abundance of elements. "SolarSystemAbundances." Image by MHz`as, 2012, Wikimedia Commons. CC-BY-SA 3.0 license. Numerical data from Katharina Lodders, "Solar System Abundances and Condensation Temperatures of the Elements," *The Astrophysical Journal* 591, no. 2 (2003).

INDEX